Assessment of Pesticide Use Reduction Strategies for Thai Highland Agriculture

DEVELOPMENT ECONOMICS AND POLICY

Series edited by Franz Heidhues, Joachim von Braun and Manfred Zeller

Vol. 73

PL ACADEMIC RESEARCH

Assessment of Pesticide Use Reduction Strategies for Thai Highland Agriculture

Combining Econometrics and Agent-based Modelling

Christian C. W. Grovermann

PL ACADEMIC RESEARCH

Bibliographic Information published by the Deutsche Nationalbibliothek
The Deutsche Nationalbibliothek lists this publication in the Deutsche Nationalbibliografie; detailed bibliographic data is available in the internet at http://dnb.d-nb.de.

Zugl.: Hohenheim, Univ., Diss., 2014

Library of Congress Cataloging-in-Publication Data
Grovermann, Christian C. W. (Christian Cornelius Wilhelm), 1982- author.
Assessment of pesticide use reduction strategies for Thai highland agriculture : combining econometrics and agent-based modelling / Christian C. W. Grovermann. — First edition.
pages cm. — (Development economics and policy, ISSN 0948-1338 ; vol. 73)
ISBN 978-3-631-65784-3
1. Pesticides—Thailand—Econometric models. 2. Pests—Integrated control—Thailand—Econometric models. I. Title. II. Series: Development economics and policy ; Bd. 73.
SB950.3.T5G76 2015
628.1'684209593—dc23

2014049517

D 100
ISSN 0948-1338
ISBN 978-3-631-65784-3 (Print)
E-ISBN 978-3-653-05134-6 (E-Book)
DOI 10.3726/978-3-653-05134-6

PL Academic Research is an Imprint of Peter Lang GmbH.
Peter Lang – Frankfurt am Main · Bern · Bruxelles · New York · Oxford · Warszawa · Wien

This publication has been peer reviewed.

www.peterlang.com

Acknowledgements

First of all, I am very grateful to my supervisors at the Department of Land Use Economics in the Tropics and Subtropics, Hohenheim University: Prof. Dr. Thomas Berger and Dr. Pepijn Schreinemachers (who now works at the World Vegetable Centre in Taiwan). Their critical feedback, encouragement and technical skills greatly supported my research, and I was able to learn a lot from them regarding research content and methodology, as well as scientific rigour in general, all of which will be of great use to me in the future. They also supported me to receive funding from the Collaborative Research Centre G1 project, 'The Uplands Program' (SFB 564) and to spent time at the project's field office in Thailand. In this regard, I would also like to acknowledge the Deutsche Forschungsgemeinschaft (DFG), which financed SFB 564, as well as the Hermann and Edith-Karla Eiselen Commemoration Foundation, which supported the publication of my dissertation through a grant.

Being part of the Collaborative Research Centre not only provided me with the necessary funding, but also gave me an opportunity to gain experience as part of a large-scale, interdisciplinary scientific project, one located at the nexus of applied and fundamental research. It also allowed me to gain important insights into field research, as well as collaborative work and state-of-the-art science. I am very thankful to my second official thesis supervisor, Prof. Dr. Dr. h.c. Franz Heidhues, who was the Director of The Uplands Program for a number of years. Discussing my research with him helped me to place my study within a wider context and gave me the confidence to pursue my work. I also benefited greatly from Thai members of The Uplands Program based at Kasetsart University, who played an important role in my data collection and analysis activities, and encouraged me by showing a great deal of interest in my research. In particular, I should thank Prof. Dr. Suwanna Praneetvatakul and Dr. Chakrit Potchanasin. The farmers I studied in the Mae Sa watershed and in Doi Angkhang in northern Thailand also deserve a special mention here, since they patiently provided large amounts of data without receiving any direct benefits in return.

Moreover, I would like to thank my girlfriend and colleague Oom, who supported me a lot during the three years of my research project, both mentally and practically, for example by making sense of the interviews I held with IPM farmers in Doi Angkhang with her translations and explanations. I must also thank my other colleagues at The Uplands Program and at Hohenheim University for the many lively discussions I had with them, for sharing their knowledge with me and for simply sharing good times together. I was very fortunate to have very pleasant, intelligent and entertaining company during my Ph.D. project.

Last but not least, I would like to express special thanks to my parents, who during my research were always there for me with a sympathetic ear, cheering me on during the difficult stages of my Ph.D. research and writing, and relieving me of the many everyday burdens I would normally have endured during such a stressful time.

Summary

The rapid growth in pesticide use has become a significant problem for Thailand, as it has in many other developing countries with intensifying agricultural activities. Risks to farmers' health during application and picking, the exposure of consumers to chemical residues on food, as well as the contamination of water bodies and the decline in ecosystem resilience, are among the consequences of heavy pesticide use in Thailand. Especially in the north of the country, the adoption of high-value fruit and vegetable crops has caused farmers to heavily rely on agrochemicals, with the use of highly toxic pesticides now being widespread.

For several decades policy-makers in Thailand have provided an environment favourable to the spread of synthetic pesticide use. Countervailing efforts aimed at the diffusion of Integrated Pest Management (IPM) activities since the 1990s and, more recently, the certification of Good Agricultural Practices (GAP), have so far had a limited impact. With increased awareness of the negative consequences of heavy pesticide use and its reduction firmly on the political agenda nowadays, more information on the economic conditions needed to reduce the use of pesticides and the policies needed to do so is required.

This research study attempts to fill this knowledge gap and generate new insights on pesticide use, based on the following three objectives. First, it aims to quantify pesticide productivity and pesticide overuse, second, it explores determinants of the adoption of sustainability innovations, and, most importantly, it evaluates the adoption of IPM in combination with a series of market-based instruments that can boost the transition to more sustainable pest control practices, such as taxes and subsidies. Econometric approaches lend themselves to the analysis needed to meet the first two objectives. Meanwhile, for the third objective an ex-ante assessment of the results of a simulation model was deemed a suitable approach to take, with the Mathematical Programming-based Multi Agent System (MPMAS) providing an effective and comprehensive policy evaluation tool for such a purpose.

These different methods are applied here using farm- and plot-level data from one intensive upland vegetable production system in northern Thailand – the Mae Sa watershed. The results of the econometric analysis were directly integrated into the simulation model in order to account for the empirical relationship between pesticide use, pest damage and output, and in order to predict agent innovativeness, using the estimated innovation adoption determinants. Innovativeness is used to assign agents to innovation segments and, consequently, define their level of access to innovations, which in this case refers to IPM for leafy vegetables production. An MPMAS model is used to assess pesticide use reduction strategies, because it is able to give an exact representation of the innovation diffusion process. Furthermore, it adds to the complexities of previous models by simulating the response of a heterogeneous population of farm agents to the availability of sustainable alternatives and market-based policy interventions. Individual policies, or packages of policies, can thus be more accurately assessed in terms of their impact and cost-effectiveness. Also, using the model, the assessment of trade-offs between pesticide use reduction and income changes can be traced back to agent-specific characteristics and land-use. Another unique aspect of this research is the fact that it considers negative pesticide externalities when estimating levels of pesticide overuse, and this allows for the comparison of tax scenarios within the model with the predicted difference between private and social levels of optimal pesticide use.

The findings show that about 80% of the pesticide quantities applied is in excess of the social optimum, while the difference between the private and social levels of overuse is small in the case study area. Therefore, the results of the econometric analysis suggest that including pesticide externalities in the price of pesticides would only have a small effect in terms of reducing pesticide overuse. The modelling results confirm that this is true for low and medium tax levels, whereas a high *ad valorem* tax that increases pesticide prices by 50% could achieve a moderate pesticide use reduction, of 8.5% on average. However, it would also decrease average incomes by 7.5%. The values for proportional taxes are similar, but due to their design they can somewhat better target highly toxic pesticide use, making them the preferred option. The pesticide substitution options shown by the MPMAS model reveal that the response to a tax is more tangible than the econometric analysis implies; nevertheless, the simulated reductions achieved with a high stand-alone tax fall short of the desired reduction targets. The main benefit of introducing a tax rests upon the generation of revenues that can be reinvested. In this context, the payment of direct compensation based on farm size is evaluated alongside a series of policies to promote IPM. A high proportional tax in combination with IPM access, and a 60% to 80% bio-pesticide subsidy, is shown to be the smartest policy package. Substantial pesticide use reductions ranging from 24% to 34% can be achieved over a five year period, with a cost-effectiveness that is clearly superior to lump sum compensation payments, price premiums for IPM produce, or transitional payments for land under IPM. In the case of Thailand, the practicability and ease of implementation of a bio-pesticide subsidy is provided by the existing infrastructure for input subsidies. The profitability of IPM is instrumental in its diffusion, as the model shows, within five years 45% of agents adopt IPM on a proportion of their land in the case of a policy package involving a 60% bio-pesticide subsidy, while 75% of agents adopt in the case of a package involving an 80% bio-pesticide subsidy. The area of cultivated land under IPM in the watershed is 22% and 38% respectively. Income losses due to tax are made up for and exceeded by income gains from the adoption of IPM, but net gains are not evenly distributed; agents polluting less generally benefit more than highly polluting agents.

The model outcomes demonstrate that it is possible to bring down levels of pesticide use significantly without income trade-offs, by combining the right policy instruments with the availability of IPM. These results are based on the adoption of IPM for leafy vegetables only; the availability of IPM for other cash crops would most likely enhance the observed effects. From this case study, it becomes clear that policy-makers should not shy away from making pesticides more expensive using a tax. If the resulting revenues are used to create financial incentives and support rural advisory services that prioritise the promotion of IPM, on-farm practices could actually be changed, which is something that voluntary instruments have failed to deliver so far. It is shown here that knowledge transfer, from extension services to farmers, as well as farmer to farmer communications, is crucial in facilitating the adoption of sustainability innovations.

Zusammenfassung

Der rasante Anstieg des Pestizideinsatzes stellt ein erhebliches Problem für Thailand, wie auch für andere Entwicklungsländer mit einer sich intensivierenden Landwirtschaft, dar. Zu den negativen Auswirkungen der starken Nutzung von Pestiziden in Thailand zählen unter Anderem Risiken für Bauern während der Ausbringung der Spritzmittel und der Ernte, Risiken für Konsumenten durch Pestizidrückstände auf Lebensmitteln, sowie die Belastung von Gewässern und eine Beeinträchtigung der Widerstandsfähigkeit gesamter Ökosysteme. Besonders im Norden des Landes hat die Einführung ertragreicher zum Verkauf bestimmter Feldfrüchte dazu beigetragen, dass Bauern in großem Maße auf Agrochemikalien angewiesen sind, wobei die Nutzung toxischer Substanzen weit verbreitet ist.

Über viele Jahrzehnte hinweg haben Politiker die Ausbreitung synthetischer Pflanzenschutzmittel begünstigt. Gegenläufige Bemühungen, die seit den 1990er Jahren auf die Verbreitung integrierter Schädlingsbekämpfungsmaßnahmen (IPM) oder, in neuerer Zeit, auf die Zertifizierung guter Praktiken im Landbau (GAP) abzielen, hatten bisher einen eher geringen Wirkungsgrad. Das gestiegene Bewusstsein für die negativen Auswirkungen des massiven Pestizideinsatzes und der politische Wille diesen zu reduzieren, sorgen für einen Bedarf an mehr Informationen über die wirtschaftlichen Rahmenbedingungen und geeignete Reduzierungsmaßnahmen.

Diese Forschungsarbeit versucht die bestehende Wissenslücke zu schließen und neue Erkenntnisse in Bezug auf die folgenden drei Ziele zu schaffen: Die Quantifizierung der Pestizidproduktivität und -übernutzung, die Untersuchung der bestimmenden Faktoren der Übernahme von Nachhaltigkeitsinnovationen und insbesondere, die Evaluierung dieser Übernahme in Kombination mit einer Reihe marktbasierter Instrumente, welche den Übergang zu einer nachhaltigeren Schädlingsbekämpfung fördern sollen, wie z.B. Steuern oder Subventionen. Für die Analyse der ersten beide Ziele bieten sich ökonometrische Ansätze an, für das dritte Ziel ein Simulationsmodell. Dafür eignet sich besonders das Agenten-basierte Modell MPMAS, womit Politikmaßnahmen umfassend bewertet werden können.

Grundlage der Forschungsarbeit sind Daten, welche auf Farm- und Plotebene für ein Gebiet im Bergland Nordthailands, dem Mae Sa Einzugsgebiet, erhoben wurden. In dieser Region wird intensiver Gemüseanbau praktiziert. Die Ergebnisse der ökonometrischen Analyse wurden direkt in das Simulationsmodell integriert um die empirische Beziehung zwischen Pestizideinsatz und Ertrag, sowie die geschätzten Faktoren der Übernahme von Innovationen für die Vorhersage der Innovationsfähigkeit der Agenten zu berücksichtigen. Die Zugehörigkeit von Agenten zu Innovationssegmenten im Modell wird durch deren Innovationsfähigkeit bestimmt, was wiederum den Zugang zu Innovationen regelt, in diesem Fall insbesondere zu IPM für Blattgemüse. Die vorliegende MPMAS-Anwendung zur Bewertung von Strategien der Reduzierung des Pestizideinsatzes in Thailand zeichnet sich durch die genaue Darstellung des Prozesses der Innovationsdiffusion aus. Außerdem konnte die Komplexität bisheriger Modellierungsansätze in diesem Bereich gesteigert werden, indem die Reaktionen einer heterogenen Agentenpopulation auf die Verfügbarkeit von nachhaltigen Anbaualternativen und auf marktbasierte Politikinterventionen simuliert wurden. Die Auswirkungen und die Kosteneffektivität einzelner oder gebündelter Maßnahmen können damit mit größerer Realitätsnähe untersucht werden. Um den Zielkonflikt zwischen einer Reduzierung der Pestizidnutzung und Einkommensveränderungen besser zu bewerten, kann auf agentenspezifische Eigenschaften und Landnutzungen zurückgegriffen werden. Eine weitere Neuerung dieser Forschungsarbeit besteht in der Berücksichtigung der negativen Externalitäten bei der Bestimmung der Pestizidübernutzung. Dadurch kann der Effekt der im Modell simulierten Pestizidsteuern mit dem Unterschied zwischen privatem und gesellschaftlichem Optimum des Einsatzes von Pestitiziden verglichen werden.

Die Resultate zeigen, dass etwa 80% der ausgebrachten Pestizidmenge über dem gesellschaftlichen Optimum liegt, wobei für die vorliegende Fallstudie der Unterschied zwischen dem privatem und dem gesellschaftlichen Übernutzungsniveau gering ist. Die Ergebnisse der ökonometrischen Analyse deuten daher darauf hin, dass die Internalisierung der externen Kosten nur eine geringe Auswirkung auf die Reduzierung der Pestizidübernutzung hätte. Die Modellergebnisse bestätigen dies für eine geringe und mittlere Steuerbelastung, wohingegen eine stärker belastende Mehrwertsteuer in der Höhe von 50% eine mäßige Reduzierung des Pestidideinsatzes um durchschnittlich 8,5% erreicht. Jedoch fallen auch die Einkommen im Durschnitt um 7,5%. Die Werte für eine proportionale Steuer sind ähnlich, wobei diese auf Grund der Gestaltung etwas genauer auf toxische Pestizide ausgerichtet ist, was Sie als zu bevorzugende Maßnahme erscheinen lässt. Die Pestizidsubstitutionsmöglichkeiten im MPMAS-Modell sind dafür verantwortlich, dass die Reaktion auf die Steuer greifbarer ausfällt als die ökonometrische Analyse impliziert. Nichtsdestotrotz bleibt die simulierte Reduktion hinter politisch wünschenswerten Zielen zurück, wenn eine Steuer als alleinige Maßnahme eingeführt wird. Der Hauptnutzen der Steuer besteht darin, dass zusätzliche Staatseinnahmen erzeugt werden, die reinvestiert werden können. In diesem Zusammenhang wurde eine flächenbezogene Kompensationszahlung nebst einer Reihe von IPM Fördermaßnahmen evaluiert. Eine hohe proportionale Steuer zusammen mit IPM Zugang und einer 60–80%igen Subvention von Biopestiziden erwies sich auf der Grundlage der Simulationsergebnisse als sinnvollstes Maßnahmenbündel. Ein erheblicher Rückgang des Pestizideinsatzes um 24–34% konnte damit in einem Zeitraum von 5 Jahren erreicht werden, wobei die Kosteneffektivität klar diejenige von Preisaufschlägen für IPM Erzeugnisse oder von IPM-flächenbasierten Übergangszahlungen übersteigt. Für Thailand ist zudem die Umsetzbarkeit durch das bestehende System der Inputsubventionen gewährleistet. The Rentabilität des mit IPM angebauten Gemüses trägt maßgeblich zu dessen Verbreitung im Modell bei. Innerhalb von 5 Jahren wird IPM von 45% der Agenten auf einem Teil ihres Betriebs übernommen, wenn das Maßnahmenbündel eine 60%ige Biopestizidsubvention enthält, und von bis zu 75% der Agenten wenn die Subvention im Maßnahmenbündel 80% beträgt. Mit IPM bewirtschaftetes Land liegt dann bei 13% beziehungsweise 17% des gesamten Ackerlands. Einkommenseinbußen auf Grund der Steuer werden von Einkommenszugewinnen durch die Übernahme von IPM Aktivitäten ausgeglichen oder übertroffen. Nettogewinne sind aber in der Agentenpopulation ungleich verteilt, da Agenten, die weniger verschmutzen, in stärkerem Maße profitieren als jene, die der Umwelt stärker schaden.

Die Modellergebnisse weisen darauf hin, dass eine bedeutsame Verringerung der Pestizidnutzung ohne einen wirtschaftlichen Zielkonflikt möglich ist, wenn die richtigen Politikinstrumente mit IPM verbunden werden. Die vorliegenden Ergebnisse basieren ausschließlich auf der Übernahme von IPM für Blattgemüse. Es ist davon auszugehen, dass die Verfügbarkeit von IPM für andere Gemüsearten die beobachteten Auswirkungen verstärken würde. Aus dieser Fallstudie wird ersichtlich, dass Entscheidungsträger sich nicht davor scheuen sollten den Preis von Pestiziden durch eine Steuer zu erhöhen. Wenn die sich daraus ergebenden Einnahmen sinnvoll für die Finanzierung von IPM-Anreizen eingesetzt werden und ländliche Beratungsorganisationen die Förderung von IPM zur Priorität machen, kann sich die landwirtschaftliche Praxis tatsächlich ändern, was bisher durch freiwillige Maßnahmen nicht geschehen ist. Es wurde gezeigt, dass Wissenstransfer von Beratungsorganisationen zu den Bauern, sowie die Kommunikation zwischen den Bauern, wichtige Faktoren bei der Übernahme von Nachhaltigkeitsinnovationen sind.

Table of Contents

List of Tables

List of Figures

Abbreviations

ABM	Agent-based Modelling
DANIDA	Danish International Development Agency
DOA	Department of Agriculture (Thai Ministry of Agriculture & Cooperatives)
DOAE	Department of Agricultural Extension (Thai Ministry of Agriculture & Cooperatives)
EIQ	Environmental Impact Quotient
FAO	Food and Agriculture Organisation of the United Nations
FFS	Farmer Field Schools
GAP	Good Agricultural Practices (Referring to the public Q-GAP standard in Thailand)
Hh	Household
IPM	Integrated Pest Management
LUCC	Land-use and Land Cover Change
MILP	Mixed Integer Linear Programming
MP	Mathematical Programming
MPMAS	Mathematical Programming-based Multi-Agent System
NGO	Non-Governmental Organisation
NOEC	No Observed Effect Concentration
OECD	Organisation for Economic Cooperation and Development
PEA	Pesticide Environmental Accounting
PNEC	Predicted No Effect Concentration
RHS	Right-Hand Side
SPE	Solid Phase Extraction
SWAT	Soil and Water
TEC	Total External Cost
USD	United States Dollars
WHO	World Health Organisation

1. Introduction

1.1 Problem statement

Agricultural commercialisation is a consequence of economic growth and urbanisation (Pingali, 2001). Economic development withdraws labour from the agricultural sector and decreases the availability of land, while it also provides better infrastructure and places new production technologies at the disposal of rural communities. Simultaneously, demand for food, especially in urban centres, rises. In this context, the intensification of crop production poses a continuing challenge to many low- and middle-income countries, particularly as such intensification is seen as having a major and detrimental impact on terrestrial and aquatic ecosystems (Tilman, 1999).

The process of agricultural intensification is often accompanied by problems of pesticide overuse and misuse (Ecobichon, 2001; Schreinemachers and Tipraqsa, 2012). Farmers and consumers in developing countries are particularly vulnerable to the health risks posed by pesticides, and especially acute poisoning (Atreya, 2008; Snelder et al., 2008; Thapinta and Hudak, 2000), because of the hazards involved when applying pesticides and the lack of knowledge that exists among farmers as to the safe and correct use of pesticides. It has also been shown that pesticides accumulate in soils, water and the food chain (Sangchan et al., 2012; Thapinta and Hudak, 2000). Surface flows and the erosion of soils that have absorbed chemical substances result in the transport of pesticides away from application sites, while heavy rainfall events can cause high concentrations of active ingredients to accumulate in water bodies, posing a serious threat to aquatic organisms (Sangchan et al., 2013). Humans are also affected by contaminated water, and particularly by pesticide residues left on agricultural produce. Food safety is an important issue for agricultural exports, but has also gained in importance within domestic markets in developing countries (Amekawa, 2009; Unnevehr, 2000). Indirectly, pesticides threaten the resilience and long-term productivity of ecosystems by disrupting natural pest control processes. As natural predators disappear and pests become resistant, the application of pesticides needs to increase and so expenditures rise. Despite high externalities and increasing input costs, farmers continue to use pesticides due to the perceived high withdrawal costs of having to give up their existing chemical pest control practices (Wilson and Tisdell, 2001). Policies in developing countries do not adequately address pesticide risks, as policy-makers often fear that disincentivising or restricting pesticide use will harm food production and, as a consequence, have a negative impact on farm household incomes (Carvalho, 2006). In fact, to advance the intensification of agriculture, many developing countries have rules in place that give farmers incentives to use more pesticides.

Besides policies, innovations have a role to play in the spread of pesticide-intensive agriculture, as they are widely regarded as the key to addressing food security and poverty reduction. Policy-makers are keen to promote agricultural innovation as being pro-poor and sustainable at the same time; however, economic rural development and environmentally sustainable production can have conflicting goals, especially in the short run. It is possible to lift farmers out of poverty by introducing high-yielding cash crop varieties and by providing them with access to markets. In Asia, the innovations introduced by the Green Revolution, such as improved seeds, mineral fertilizers and chemical pesticides, have spread extensively, which has improved the livelihoods of many farmers (Hazell, 2009; Zeller et al., 2013). It has also changed farmers' key objective, from that of food self-sufficiency to one of profit maximization. Inputs are mainly traded and no longer generated by households, while the product mix has shifted from diverse to highly specialised (Pingali and Rosegrant, 1995). The consequences of this rapid commercialisation process on land-use are mono-cropping and a drastic shortening of fallow periods. The ecological flip-side of productivity enhancing innovations is manifested in high levels of soil degradation, as well as increased weed and pest pressure (Pingali, 2001). The abandonment of agro-ecological practices, such as rotations, and the spatial and temporal decrease in structural diversity, has led to a higher incidence of disease and pests. This can be observed despite the increasing use of chemical pesticides. Advances in breeding disease- and pest-resistant crops, as well as the effectiveness of pesticides, are often transient, since pests adapt and can resurge in large numbers (Tilman, 1999; Tilman et al., 2002). A loss of natural predators in intensively used and simplified agro-ecosystems can also exacerbate pest problems (Tscharntke et al., 2005; Wilby and Thomas, 2002).

Like other emerging economies with an export-oriented agricultural sector, Thailand has very rapidly increased its agricultural pesticide use in recent years (Schreinemachers and Tipraqsa, 2012). Whereas Thai farmers used 1.2 kg of active pesticide compounds per hectare in 1997, by 2010 they were using 3.7 kg/ha – an average increase of 9% per year (Praneetvatakul et al., 2013). This increase is reflected by a surge in the import of fungicides, insecticides and herbicides in particular, as shown in Figure 1. Against the backdrop of increasingly heavy pesticide use in Thailand, Praneetvatakul et al. (2013) used the Pesticide Environmental Accounting (PEA) tool to estimate the external costs of pesticide application. They applied the tool to the country as a whole, as well

Figure 1: Increase in pesticide imports into Thailand

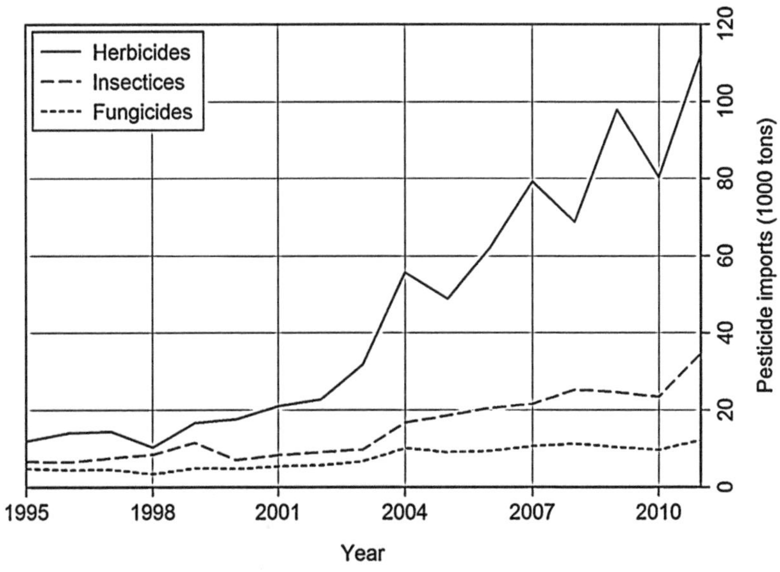

(Source: Office of Agricultural Economics, Thailand)

as to two distinct agricultural systems: rice and intensive upland horticulture. They estimated the negative externalities (so-called average external costs) to be USD 27/ha for Thailand as a whole, USD 19/ha using a dataset of 224 rice farmers, and USD 106/ha using a dataset of 295 farmers practicing intensive upland horticulture. Meanwhile, for the US and Germany, external costs have been estimated to be USD 78.5/ha and USD 113.8/ha respectively (Pretty et al., 2001). While higher income countries generally spray more pesticides per hectare, the broad-spectrum pesticides used in lower income countries are often more hazardous (Carvalho, 2006; Konradsen et al., 2003). At the same time, clean up-costs per unit of applied pesticide are lower in these countries (Leach and Mumford, 2008). Overall, the external cost estimates for Thailand, especially for horticulture, are an indication of severe market failure.

Thai policy-makers have been quite supportive of pesticide use, offering cheap credit to buy inputs, giving tax exemptions for agricultural pesticide imports, and facilitating the free distribution of pesticides during major pest outbreaks (Praneetvatakul et al., 2013). Efforts have been undertaken to limit pesticide use by restricting the import of highly hazardous pesticides, while at the same time trying to reduce pesticide demand by promoting organic agriculture, running farmer field-schools on Integrated Pest Management (IPM), and introducing a public certification programme based on Good Agricultural Practices (GAP) (Schreinemachers et al., 2012). However, the effect of these efforts has been limited. Uncertainty over the impacts of pesticide use reduction strategies is a major obstacle to implementing more effective policies, and to shifting the focus from supporting conventional crop protection practices to widely promoting the adoption of IPM innovations.

1.2 State of the art and research gaps

Concerns that restrictions on pesticide use will put food production and food security at risk are not usually based on empirical analysis. In fact, there are very few analytical tools available for making such an assessment, and scientific studies on the impact of different policy instruments on farm income, pesticide use and the environment have remained relatively few and far between (Skevas et al., 2012). The gravity of the problems associated with heavy pesticide use, especially in emerging economies, call for further well-grounded and science-based information to be generated. If taking the concept of sustainable intensification seriously, it is important to consider low-input, integrated farming techniques alongside direct interventions such as taxes. However, the innovation diffusion and adoption processes for sustainability innovations have so far received little attention in the context of pesticide policy. Last but not least, a comprehensive assessment of pesticide policies presupposes a good understanding of pesticide productivity and optimal application amounts.

1.2.1 Optimal pesticide use and pesticide overuse

The common approach used to quantify optimal levels of pesticide use involves the estimation of an agricultural production function in which pesticides are included as a damage control agent (Lichtenberg and Zilberman, 1986). Scholars generally agree that the use of a damage control approach is necessary to avoid overestimating the marginal effect of pesticides. The exponential specification of the damage abatement term within this function has given realistic results in a number of crop protection studies (Jah and Regmi, 2009; Pemsl et al., 2005; Praneetvatakul et al., 2003). These studies found that pesticide productivity is low in developing countries such as Nepal, China and Thailand. Production functions are generally estimated from farm-level data. They therefore, give an estimate of the optimum level of pesticide use from the point of view of a farmer (that is, a "private" optimum). However, the use of pesticides also creates negative externalities, such as adverse health effects among farm workers and consumers, as well as imbalances in the functioning of ecosystems. As the external effects are not transmitted to farmers through the prices of pesticides, the private optimum level of pesticide use is likely to be in excess of what is optimal from a societal point of view (Pretty et al., 2001). In other words, the "social" optimum would be below the private optimum if negative pesticide externalities were to be included. This social optimum level of pesticide use can be estimated by adding the external costs of pesticides to their purchase costs. Yet, no studies have previously done this, because there was no method available to quantify the external cost of an individual farmer's pesticide application.

Economic analysis of the external costs related to pesticide use has been carried out at the national level for the USA (Pimentel, 2005; Pimentel et al., 1993), UK (Pretty et al., 2000) and Thailand (Jungbluth, 1996), estimating the combined cost of pesticides. Leach and Mumford (2008, 2011) recently developed a method which allows a researcher to determine the externalities associated with individual pesticides, called the Pesticide Environmental Accounting (PEA) tool. The PEA tool allows for pesticide externalities to be disaggregated down to the field or farm level. The novelty of this study is that it combined the PEA tool with a common production function framework, to determine marginal costs and the benefits derived from pesticides. The results serve as an important indication of how much pesticides are overused from a societal point of view, and also how responsive farmers' behaviours might be to increases in the price of pesticides through the application of a tax for example. A thorough evaluation of the effect of pesticide use reduction policies requires the modelling of land-use decision-making to take place under resource and policy constraints, which is why the analysis in this study is complemented by use of the agent-based model MPMAS, as further explained below. The estimation of a production function including a damage control term, is important for the specification of the model used here, and the results of the overuse analysis are essential, to put the model's simulation outcomes into context.

1.2.2 Diffusion and adoption of innovations to reduce pesticide use

Innovation diffusion research is relevant in many disciplines, such as politics (Granovetter and Chung, 1997), epidemiology, geography and sociology (Valente, 2012). The theory of innovation diffusion, as formulated by Rogers (2003), explains how a technology spreads though an innovation network, such as a farming community for example. Within the theory, the adoption of an innovation is conceived of as a process whereby innovative members of the network adopt earlier than cautious members. The innovativeness of an individual, that is, his or her likelihood to adopt, is a function of personal network characteristics, as well as other characteristics such as education (Valente, 2005). Network characteristics refer to the concepts of the cosmopoliteness, exposure and centrality of an individual (Valente, 2012). The number of ties actors have with the outside world describes their cosmopoliteness, while inside the network the ratio of actors' contacts that know about an innovation to those that do not know determines the level of exposure. Lastly, centrality refers to a measure of importance or prominence within a network. These three variables can be quantified for each member of a network, as is the case for other characteristics, such as education, age and resources.

In practice, the innovativeness of farmers in a rural innovation network can be quantified by econometrically estimating adoption probabilities (Schreinemachers et al., 2009). For instance, the adoption of agro-forestry in Nepal has been estimated using logistic regression analysis (Neupane et al., 2002), while the adoption of new pigeon pea varieties under seed access constraints was modelled using the double hurdle method by Shiferaw et al. (2008). Wollni and Zeller (2007) examined what determines participation in specialty coffee markets in Costa Rica. The conceptual framework of agricultural innovation adoption used by Abadi Ghadim and Pannell (1999) concluded that distance to and contact with other adopters of an innovation are important factors. However, the above mentioned studies only partly included network characteristics in their regression models. Contacts with extension or producer organizations were mostly captured, but farmers' exposure levels were often neglected, since data was not always available. Also, many econometric studies are static and focus on simply analysing the determinants of adoption, while neglecting the patterns of innovation diffusion; they also do not provide information on when adoption takes place (Schreinemachers et al., 2009). The modelling of network dynamics can provide important insights into the temporal dimension of adoption; for example, an agent-based spatial model such as MPMAS can consider the diffusion of an innovation when simulating land-use decision-making over time (Berger, 2001).

So far, in the context of the debate on restricting pesticide use in tropical countries, little is known about the determinants of farmers' innovativeness or lack thereof, nor about the diffusion process for pesticide use reduction innovations. A greater level of knowledge regarding the adoption of sustainability innovations, such as the certification for GAP and also IPM practices, is needed to better understand the opportunities and challenges involved. The Thai public GAP standard is a very relevant innovation that has been increasingly adopted by farmers in the region. Aimed at local small-scale producers, it has been shown a useful means of linking them to markets (Amekawa, 2013), but very little is known about the adoption process itself. With more data available, this would provide a suitable object of investigation for future adoption research. The econometric estimation procedure for adoption determinants and probabilities can be improved by choosing the most appropriate technique, and by explicitly including network characteristics. Insights into the innovativeness of farmers can then be combined with agent-based modelling, where social network dynamics as well as profit-maximisation explain the innovation diffusion process.

1.2.3 Assessment of pesticide use reductions

Two important policy appraisals of pesticide use reduction interventions have been carried out using mathematical programming models. First, Falconer and Hodge (2000) developed a case-study farm model for the UK, to evaluate low-input farming practices in combination with four distinct pesticide taxation and levy schemes. Linking an economic model – used to simulate agricultural production decisions – with ecological indicators, they found significant trade-offs between economic and environmental objectives. The authors noted that the level of a pesticide tax needs to be set at a high enough level if the taxation mechanism is to be effective. The topic of pesticide use reduction policies was also addressed by Jacquet et al. (2011), using an economic model set at the national level within the French agricultural sector. This model suggested that taxation would help reduce pesticide use considerably, but gave no evidence of significant income losses to farmers, as long as integrated farming techniques were widely adopted.

Besides MP models, econometric analysis has produced some insights into the impact of pesticide policies. This type of analysis focuses more on pesticide productivity and thus the response of pesticide use to policies, while the impact such policies have on farm household incomes takes a back seat. Skevas et al. (2012) studied the effects of pesticide use reduction policies on Dutch cash crop production. The object of their analysis was to assess the effectiveness of taxes, subsidies and quotas on pesticide use, as well as on the pesticides' environmental spill-overs, such as on farm biodiversity levels. The findings of this study suggest that even extremely high taxes and penalties result in only small reductions in pesticide use and thus spill-overs, since the demand for pesticides is very price inelastic. Own-price elasticity is the ratio of the percentage change in the quantity of a pesticide that farmers acquire, to the percentage change in the pesticide's price, while cross-price elasticity represents the ratio of the percentage change in the quantity of a pesticide that farmers acquire, to the percentage change in the price of another pesticide. Reviewing the literature on pesticide demand and examining the effects of an environmental tax on pesticide use in Mexico, Pina and Forcada (2004) showed that generally the demand for pesticides is price inelastic, which is in line with the findings of Skevas et al. (2012). In contrast, the results of Agne (2000) suggested that pesticide use within the coffee production sector in Costa Rica is rather price elastic. Analysing the income effects of a pesticide tax on coffee production using a partial budget model, Agne found that the income effects of applying a pesticide tax were not substantial. The partial budget approach is, however, limited, since it is based on a fixed technology package and does not take into account factor substitution.

As the above mentioned studies illustrate, analytical approaches vary in terms of scope and methodology, which may help explain why so far there has been little or no consensus on the exact impacts of pesticide use reduction policies. One can ascertain that, especially in the context of tropical agriculture, to date little attention has been paid to investigating the implications of pesticide use reduction policies on land-use, the environment and smallholder livelihoods in a comprehensive manner. Because the level of understanding on policy impacts remains sketchy, it is difficult to give sound advice to decision-makers, and in the absence of any scientific analysis of the exact costs and benefits associated with pesticide use reduction, debates about pesticide policies in these countries remain prone to the influence of ideology and commercial interests. The spatial, social and economic separation of urban-based interest groups, such as chemical manufacturers and food retailers, and the rural growers actually using the pesticides, means that the benefits and risks have also remained separated (Kunstadter, 2007). Therefore more research-based information on the feasibility of policy interventions and alternative crop protection methods is required in order to allow for their successful implementation, and to balance opposing interests. The intention of IPM, as an innovation that permits the sustainable intensification of crop production and pesticide risk reduction, is to be pro-poor and environmentally sound at the same time. However, uncertainty still exists on its actual economic impact at the farm level, an important factor preventing investment in the knowledge transfer needed to generate higher adoption rates.

Falconer and Hodge (2001) pointed out that simple economic models of pest management decisions can be unrepresentative or even misleading if used as a basis for policy recommendations. The following paragraph explains how inadequate they consider representative farm models to be: "Models are useful tools for analysing production and ecological changes under alternative policy scenarios. However, the use of the representative farm approach clearly cannot represent the diversity of farm types and the behaviour of individual farmers; the case-study results provide a broad guide only. The spatial and temporal distributions of changes, and their implications for environmental quality were outside the scope of such analysis. Investigation of this is a priority area for further work; it is also important to examine the potential consequences of policy intervention in different production contexts." (Falconer, 2000)

Many of the weaknesses mentioned above can be remedied through the use of multi-agent systems. Using these systems, a range of diverse farm types and individual behaviours can be modelled in space and time, which is a useful addition to the previous studies on crop protection policies. To assess the impact of different pesticide use reduction policies in the Thailand context, this study uses the Mathematical Programming-based Multi Agent Systems (MPMAS) model (Berger, 2001; Berger et al., 2006). MPMAS has been developed and widely tested to perform ex-ante assessments of changes in policies and the introduction of new technologies in developing countries (Abadi Ghadim and Pannell, 1999; Marohn et al., 2013; Quang et al., 2014; Schreinemachers et al., 2007a), including in Thailand (Schreinemachers et al., 2009; Schreinemachers et al., 2010). The MPMAS application used in this study was parameterized with farm and plot level data taken from intensive and diverse production systems in the mountainous north of Thailand, and in particular the Mae Sa watershed, where the adoption of cash crops has been accompanied by very high levels of pesticide use. The model allows exploring the diffusion of policy-driven innovations among the farm population, as well as the effects of policy interventions on heterogeneous farm households and the landscape over time. It is also based on empirically-estimated damage control functions, the aim being to capture different levels of pesticide productivity. Good IPM practices for leafy vegetables were included in the model using data collected from farmers who had received extension services training and whose productions is closely monitored. In the model, econometrically estimated adoption probabilities are used to assign innovativeness scores to agents, and to simulate the diffusion of IPM. The adoption of IPM is assessed in combination with tax instruments and various other incentives, such as bio-pesticide subsidies, area payments and price premiums. This study aims to provide useful insights into the effects of a wide range of pesticide reduction strategies on a farm household population, by evaluating interventions in terms of their impact on pesticide use, land use and farm incomes. It also addresses important knowledge gaps, while also extending previous assessment methodologies in terms of scope and modelled complexity.

1.3 Research objectives

As explained above, this study addresses important research gaps in pesticide economics and policy. It aims to assess pesticide use and strategies for pesticide use reduction within the intensive upland agriculture sector of Thailand, a tropical country in transition. A key part of this aim is to carry out a comprehensive and theoretically well-grounded appraisal of crop protection policies at the farm level, which can help develop a more sustainable farming system. The results of this research should also help meet demands in Thailand for more advice to be available on the practicality of different pesticide use reduction interventions. This includes aspects such as the optimal use of pesticides from a societal perspective, the economic instruments that may be used to help reduce pesticide use, the adoption of pest management innovations, as well as the pesticide regulations related to eco-toxicological thresholds and a model of pesticide concentrations in rivers in the study area. The following three objectives helped refine the overall goal of this research, and illustrate its components:

- To analyse pesticide productivity, optimal pesticide use and pesticide overuse from a private and a social perspective, using an econometric model.
- To estimate the determinants and farmer innovativeness in relation to the adoption of an existing pesticide use reduction innovation (GAP certification), using an econometric model.
- To do an *ex-ante* assessment of different pesticide use reduction policies as well as the adoption of a newly introduced pesticide use reduction innovation, i.e. IPM, using the agent-based spatial simulation model MPMAS.

The first objective of this study is to show how to quantify pesticide overuse from a societal as well as private point of view, by combining the Pesticide Environmental Accounting (PEA) tool with a production function using a damage control specification, and then to illustrate this method using data from a horticultural production system in northern Thailand. This analysis will deliver important information on pesticide productivity, the magnitude of pesticide overuse and the elasticity of pesticide use in relation to price increases by internalising the external costs of pesticide use. These are all important aspects for the further assessment of pesticide application reductions, and in particular for the design of the MPMAS model and the definition of policy scenarios.

The second objective is to provide a better understanding of how pesticide use reduction innovations may be adopted. The results shown here do this by analysing the determinants of adoption with regard to GAP certification, and by estimating the innovativeness of farmers in the study area. In this regard, particular emphasis will be placed in this study on the inclusion of personal network characteristics as explanatory variables. The specification of the econometric model used here to estimate adoption probabilities with regard to GAP certification was an important input to the agent-based model. In MPMAS, estimated innovativeness is instrumental in helping to classify agents into innovation adopter groups, and for modelling the diffusion and adoption of IPM.

The last and key objective of this study is to provide a fully-fledged impact assessment of reduced pesticide use through a range of market-based policy interventions, as well as through the diffusion of integrated pest management practices. This objective builds on the results of the other two components of the thesis, and so combines insights gained from the econometric analysis with the capabilities of the simulation model. This third element represents the most comprehensive part of the research, and should, therefore, increase the general level of understanding that exists regarding the processes involved in, and the outcomes of, different approaches to pesticide use reduction.

Figure 2: The three research components of this study

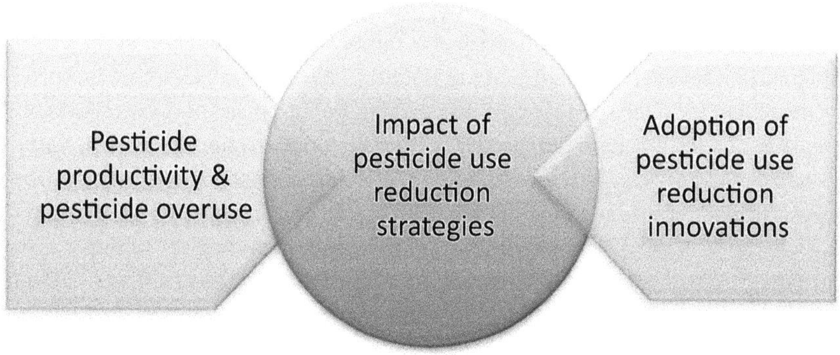

As Figure 2 shows, the three components of this research partly stand alone, but also overlap. The different approaches are tied together; in particular, the results of the econometric analysis contribute to the main component – the impact assessment. The results of the pesticide overuse analysis and the innovation adoption analysis also serve to complement the modelling outcomes and put them into perspective. Besides the three research components and respective objectives listed above, there is an overarching purpose of this study, which is of a more practical nature, i.e. the research outcomes should provide reliable evidence on which to base actual policy advice. The relevance of the topic and the demand for information require that usable, practical recommendations can be drawn from the research. This is also in line with the overall goal of the Uplands Program, the research project within whose context this research was devised. In order to make sure that the research results are sound, it is important that the selected approaches and methods are transparent and comprehensible. Consequently, the simulated impacts on the agent population outlined here will provide indications on how farmers might react to different policy interventions, something which will facilitate the design of a smart policy mix aimed at reducing pesticide use in reality.

1.4 Pesticide policy background

Reductions in pesticide use can be achieved by regulatory or so-called command-and-control interventions, by voluntary actions or through market-based instruments. Pesticide bans are the most drastic intervention, and these are required for very hazardous substances to prevent irreversible damage to the environment. Besides issues regarding enforcement and sufficient monitoring capacities, bans have been criticized as too blunt, because they do not allow farmers to deal with reduction targets that involve the gradual shift or phasing-out of pesticides which are authorized, but still of concern (Pina and Forcada, 2004). Similarly, regulations that limit pesticide use quantities are also rather blunt; pre-defined limits cannot differentiate between different types of polluters. In contrast, economic incentives or disincentives offer the prospect of reducing pollution beyond set criteria and are, from a social welfare perspective, generally more efficient. Economic incentives also encourage polluters with lower abatement costs to reduce their environmental impacts, more so than polluters with higher abatement costs (Jaeger, 2005). A wide range of market-based policy options exists, including for example taxes, subsidies, transferable permit schemes, insurance and credit instruments (Oskam, 1997). These can be applied to tackle problems of pesticide overuse and misuse. Among Organisation for Economic Cooperation and Development (OECD) countries, Denmark, Sweden, France and Norway have successfully introduced a levy on pesticides which includes some degree of differentiation according to toxicity (Pina and Forcada, 2004; Schou, 1999). Oskam (1997) evaluated how carefully designed tax or charging schemes have been useful tools, in conjunction with other measures, when wishing to address the environmental impacts of pesticides use in Europe. Effective policy interventions that attempt to mitigate the problems of pesticide overuse have for the most part been confined to high income countries. Middle income countries, such as Thailand, often shy away from taxing pesticides. However, a tax is seen as a viable policy instrument for the Thai Government, in order to complement existing legislation through market mechanisms which influence the selection of pest control measures (Praneetvatakul, 2005). However, many developing countries have rules in place that give farmers an incentive to use more pesticides, not reduce their use. Also, policies rarely address the risks and external effects of pesticides adequately, as policy-makers fear that restricting pesticide use will harm food production (Carvalho, 2006). Until recently, agricultural pesticide use in Thailand, as in other lower income countries in East and Southeast Asia, was low, and so government policies were introduced with the aim of stimulating agricultural output through the promotion of pesticide use among farmers (Praneetvatakul et al., 2013). These promotion activities have consisted of favourable loans, the free distribution of pesticides, and research funding. Such policies, as well as the pesticide industry's own marketing initiatives, have provided a favourable environment for the diffusion of chemical pest management. With increasing commercialisation business interests have emerged whose key aim is to sell seeds, fertilisers and pesticides to farmers.

The Pesticide Policy Project (Jungbluth, 1996; Praneetvatakul, 2007; Waibel et al., 1998) investigated the institutional environment and some potential pesticide use reduction strategies for several countries in Southeast Asia, and in particular Thailand. In this context, Praneetvatakul et al. (2007) traced the development of the pesticide problem in Thailand over the last 60 years, identifying institutional bottlenecks such as the rent-seeking behaviour displayed by government agencies. Big budgets for chemical pest control support were allocated against a political backdrop which favoured mono-cropping, all-year high-yielding varieties, and a belief in preventive spraying. After 1967, a policy of free pesticide distribution was in place, to control major outbreaks, but this officially ended in 1999, though parts of it are still practiced. In 1991, the government got rid of import taxes on pesticides, so in contrast to fertilisers, pesticides receive an indirect subsidy. Also in 1999, Thailand adopted the international FAO Code of Conduct on the Distribution and Use of Pesticides (FAO, 2002). Following the provisions set-out in the Code of Conduct, and in the Stockholm and Rotterdam Conventions, an international treaty dealing with chemical pesticides, the Hazardous Substances Act, was promulgated in 1992 and revised in 2005. The Act regulates the registration of pesticides in terms of manufacture, import and possession, and as a result, certain extremely hazardous compounds are now banned in Thailand, while some 298 active ingredients are officially registered. However, the slow implementation of the Act, the ease of use, and the still wide-ranging availability of broad-spectrum highly toxic pesticides, as well as overall favourable conditions (see Table 1), have prevented far-reaching changes in pest management at the farm level. Waibel (1998) provides a good overview of the observable and latent price and non-price drivers behind pesticide overuse, and this is shown in Table 1.

Table 1: Factors leading to pesticide overuse

	Price Factors	Non-price Factors
Obvious	Below market price or free distribution of pesticides through government or development organizations Subsidies for pesticide production Pesticides in credit programs Subsidies for complementary inputs Preferential rates for tax or exchange	Dominance of pesticide research Inadequate government research in environmentally benign pest management Misguided government activities in reducing pesticide damage Subsidies for pesticide intensive crops Export promotion of agricultural products
Hidden	Outbreak Budget of plant protection service Externalities of pesticide production Externalities of pesticide use	Lack of adequate definition of crop loss and pests Lack of transparency in regulatory decision making Insufficient information about pesticide risks and safer alternatives Curricula of agricultural extension and education Misinformation of farmers by chemical industry

Source: (Waibel et al., 1998)

Interviews with experts at Kasetsart University, as well as with policy-makers at the Department of Agriculture (DOA), have confirmed that the problems analysed by the Pesticide Policy Project in the 1990s and early 2000s persist today in Thailand. Alternative pest management methods that ensure the protection of valuable commercial crops are rarely applied on a large scale. The developments introduced by and achievements of the Thai government in helping to create and realise sustainability strategies are summarised by Kasem and Thapa (Kasem and Thapa, 2012), who state that there is a clear divergence between policies and actual practices. While the guidelines for more sustainable farming practices might be in place, implementation of these guidelines often lags behind.

The Royal Project was initiated in the 1960s by the King of Thailand to improve the living standards of hill farmers, to replace opium poppy cultivation and to promote a more sustainable approach to highland agriculture. Vegetable production has been one of the main activities implemented by the project, and by 2006 there were 3,000 families in 37 extension stations across five provinces in the northern part of Thailand engaged in the production of 140 kinds of vegetable crops. For these crops, the Royal Project tries to ensure that high quality standards are met (Jayamangkala, 2006). IPM has been propagated to different degrees by some of the extension stations, however, it is not generally applied.

In Thailand, starting with rice in 1976, IPM has been progressively applied to 25 other crops, such as cotton, cabbages and sugar cane (Vongsaroj, 2007). Nevertheless, IPM is a rather marginal methodology across the country as a whole. During the Green Revolution, farmers abandoned their traditional multi-cropping systems and changed to monoculture (Praneetvatakul, 2007). With high-yielding varieties, farmers could afford to also spend more on chemical pesticide inputs. However, the indiscriminate use of many such substances caused problems such as pest resurgences and resistance. The increased frequency and intensity of pest outbreaks posed an immediate threat to the gains made by the Green Revolution (Pontius et al., 2002), and in the 1990s some decision-makers realized that many farmers were trapped into using more and more chemicals (Praneetvatakul, 2007), a phenomenon that has been described as the 'pesticide treadmill' (Kenmore, 1991; Wilson and Tisdell, 2001). The influence of donor organisations, and initiatives led under the King of Thailand's name, saw some innovative IPM projects introduced alongside targeted extension work in the 1990s (Areekul, 2007). For example, the national IPM programme in Thailand has been operating since 1999, having been implemented by the DOAE in cooperation with national and international partners such as the Royal Project, the Danish Development Agency (DANIDA) and FAO (FAO, 2013b). Under the programme, initiatives such as Farmer Field Schools (FFS) have been launched to spread the use of IPM for rice and vegetable crops. Leafy vegetables are very important in terms of cultivated area and farm revenues in northern Thailand, but are also subject to high pest pressure. IPM has been used for leafy vegetables to better control the Diamond Back Moth (*Plutella xylostella*), which is a major pest for highland brassica production, causing considerable economic damage. At the Royal Project station in Doi Angkang in northern Thailand, farmers successfully practice a far-reaching agro-ecological IPM system for the growing of leafy vegetables. In other places, some alternative control methods are also being used more regularly, such as traps or bagging, but generally labour intensive practices (such as weeding), expensive materials (such as bio-pesticides), as well as limited knowledge about biological pest management practices, prevent the widespread adoption of IPM (Areekul, 2007). As a result, so far the use of IPM approaches to control pests has remained very limited. Also, supportive policies such as direct or indirect subsidies for IPM have not been introduced, and in fact the general policy environment in Thailand tends to maintain pesticide use at high levels. In 2004, the introduction of a public standard called Good Agricultural Practices (GAP) aimed to bring down pesticide use levels. While the public GAP standard has since been widely adopted, it has not been effective at reducing pesticide use levels to any great degree (Amekawa, 2013; Schreinemachers et al., 2012). In this regard, the focus on residue testing rather than changing on-farm practices and the lack of monitoring capacities have been identified as major shortcomings.

Table 2: Possible policy interventions for pesticide use reduction

Characteristics	Options	Comments on Status Quo
• **Legislation vs. budget allocation decisions** • **Central vs. local execution** • **Direct vs. indirect targeting of pesticides** • **Binding for farmers vs. voluntary participation**	Pesticide use reduction targets (e.g. based on chemicals and their environmental impact)	No binding reduction targets or usage limits Hazardous Substances Act (ban of some very toxic chemical compounds → import, production and possession illegal)
	Pesticide registration schemes	Hazardous Substances Act (specifies pesticides that need to be registered for production/import)
	Proper use regulations	Lack of work safety standards Lack of sanctions for overuse or unduly disposal
	Food safety regulations	Food Act (acceptable residue levels)
	Market-based instruments to influence the selection of pesticides	No environmental tax or subsidy (no financial support for improving access to alternative pest management) Reduced import tax for pesticides (indirect subsidy)
	Certification schemes for agricultural products	Q-GAP standard (little effect on pesticide use) Organic Thailand label (not widespread)
	Research funding (e.g. cost-benefit-risk-analysis)	Research often geared towards intensification Research results on IPM not put into practice
	Investment, extension and knowledge transfer	Substantial outbreak budget Easy access to credit for pesticides Some promotion of IPM, often initiated by external actors (FAO, DANIDA, The King, Royal Project)
	Payments for Ecosystem Services	No projects on payments for sustainable pest management practices

Source: Compiled from various sources (FAO, 2010; Jungbluth, 1996; Praneetvatakul, 2005)

After several decades of strong pesticide use growth in Thailand, the view is starting to take hold that decisions on pesticide use reductions are required at the national as well as regional level. Policy-making is evolving from a focus on promoting pesticide use, to efforts aimed at reducing it (Praneetvatakul et al., 2013). However, more information is needed to guide actual decision-making on which policies to use and which practices to promote. Table 2 shows a range of possible policy interventions available to policy-makers aimed reducing pesticide use. It also describes the characteristics which distinguish these interventions, as well as the status quo in terms of implementation in Thailand. The Hazardous Substances Act needs to be complemented by financial instruments to influence the selection of pest control measures; therefore, this research focuses on market mechanisms that are potentially relevant for Thai policy-makers. As far as such mechanisms are concerned, various studies have concluded that a tax as a stand-alone intervention is insufficient, and that a package of policy measures is required to induce farmers to change on-farm practices (Falconer, 2000; FAO, 2011; Praneetvatakul et al., 2013). Whereas taxes are considered to be rather ineffective due to the inelastic demand for pesticides in relation to prices, they can generate revenues, which can be gainfully used to promote sustainability innovations. Following the introduction, this study first tests the validity of the above statement in the context of Thailand's highland horticulture sector, and then evaluates different methods that may be used to redistribute or invest the tax revenues generated. The main criterion used for choosing the right tax rates was to find a compromise between generating sufficient government revenues and practicability. Therefore, the rates selected in the tax scenarios range between the extremely high tax rates (above 100%) found in the study of Skevas et al. (2012), and the very low tax rates (3–5%) implemented in Scandinavia since the 1990s (Schou, 1999). If tax levels are set too high, their acceptance is at risk. Pesticide taxation is already considered a highly controversial topic in Thailand, since politicians fear alienating the powerful rural electorate (Praneetvatakul et al., 2013). If, in contrast, tax levels are set too low, their effect is negligible and insufficient revenues are generated to fund alternative pest management methods, such as IPM. As far as funding for IPM is concerned, this thesis assesses three different adoption incentive instruments: Transitional area payments, price premiums for IPM produce and bio-pesticide subsidies. These are then compared to a simple redistribution of tax revenues using lump sum payments. As opposed to the adoption incentives, lump sum payments are unlikely to promote the desired change in on-farm practices. All of the redistributive or financial incentive policies considered here can, however, be used as a means to increase the acceptance of a tax.

Northern Thailand is particularly suitable for assessing the impact of pesticide reduction policies. It has seen an expansion of intensive agriculture in recent years, from the lowlands into the uplands, which, favoured by a cooler climate and more rainfall, and due to recent improvements in infrastructure, has become an important supplier of temperate and sub-tropical fruit and vegetables. The rapid adoption of cash crops by upland farmers, who until recently relied on growing rice for their own subsistence, has led to a high dependence on synthetic pesticides (Schreinemachers et al., 2011) and pesticide overuse, a trend that has been observed in many Asian countries (Pingali, 2001), and northern Thailand can be considered representative of the land use intensification that has occurred across the region as a whole. Heavy pesticide use in northern Thailand and its negative effects have been described at length (Lamers et al., 2013; Sangchan et al., 2013; Schreinemachers et al., 2011; Thapinta and Hudak, 2000), and these studies form a sound basis for the following assessment.

1.5 Structure of the thesis

This thesis continues by providing information on the material, as well as the different approaches and methods used for this research. Chapter 2 contains details on the study area, the data collection activities and the data itself. It describes in detail the characteristics of the study farms in the Mae Sa watershed and their production. Further to this in Chapter 3, the econometric analysis used in this study is explained, including how pesticide overuse is quantified, the separation between the private and social costs of pesticide use, specification of the sustainability innovation adoption model, and the prediction of innovativeness. Chapter 3 also defines the model set-up and the parameterisation of the MPMAS application, the model's interface with the econometric analysis, as well as the simulation arrangements used for the assessment of a wide range of pesticide use reduction scenarios.

The explanation of the materials and methods used is followed by the description of the model validation and results in Chapters 4 and 5. Just like the methods chapter, the results chapter also separates out econometric analysis from agent-based modelling. It first presents details of pesticide overuse and the innovation adoption results, and subsequently shows the simulation outcomes for the pesticide use reduction strategy assessment. That part of the results chapter starts with an evaluation of the stand-alone tax scenarios, then describes the tax scenarios in combination with the introduction of IPM, and continues with an appraisal of scenarios that provide innovation adoption incentives for IPM diffusion. After that, the impacts of smart policy mixes that bring together several interventions are assessed and illustrated. All policy instruments are evaluated in terms of pesticide use reductions, income changes, tax revenues and policy costs. The last section of the results chapter contains the gist of the simulations experiments and serves to formulate recommendations for policy-making. In this regard, the most practicable and cost-effective policy mixes with a significant impact are identified.

The thesis ends with a discussion of the methods and results, together with drawing several conclusions. The methodological strengths and weaknesses of the selected approaches are discussed and the methods, just like the research outcomes, are placed in the context of previous studies in the field of pesticide policy. It is also imperative to discuss the policy implications that can be derived from the present assessment of the various instruments available for pesticide use reduction. In this regard, important aspects of rural development and sustainable land use in the uplands of Southeast Asia, and the institutional environment therein, are considered. Overall, it is possible to conclude that substantial pesticide use reduction could be achieved without having to make income trade-offs.

2. Materials

2.1 Study area selection and data collection

The Mae Sa watershed area in northern Thailand was selected as the primary data collection area for the study, and was one of the two main study sites in Thailand used by The Uplands Program, a collaboration between Hohenheim University in Germany and several universities in Thailand and Vietnam. The overall aim of the project was to carry out research aimed at helping conserve natural resources and improve rural livelihoods (Heidhues and Pape, 2007; Schreinemachers et al., 2013). The Mae Sa watershed is a good example of the benefits and problems derived from agricultural commercialisation. Mountainous areas across the region have experienced a rapid intensification of agriculture activities in recent years. Favoured by a cooler climate and greater rainfall, and stimulated by recent improvements in infrastructure, upland areas such as the Mae Sa watershed have become important suppliers of temperate and sub-tropical fruit and vegetables in Thailand. The study area is located about 30 km northwest of the regional capital Chiang Mai, and is characterised by good market access and intensive upland agriculture. It covers an area of 140 km^2, with altitudes ranging from 400 m to 1,600 m above sea level (masl). The farming land around the study villages – from where the data was collected, and which are located in the central watershed and the surrounding hills – is situated at altitudes between 800 and 1,400 masl.

Since the prohibition of opium production in the late 1950s, a considerable effort has been made by the Thai government, in collaboration with international donors and the Royal Project Foundation, to replace poppy production with commercial horticulture (Heidhues and Pape, 2007). Farmers nowadays grow a wide variety of cash crops, most importantly bell peppers, tomatoes, cabbages, lettuce, onions, green beans, chayote, maize, rice, chrysanthemums, roses and litchi trees. Cropping patterns vary according to each village's location, and the elevation and slope of the land, which results in a spatially diverse agricultural land-use mix, with particular crops such as litchi being locally concentrated. However, the increase in production of high-value crops has been accompanied by heightened pest pressure and heavy pesticide use (Schreinemachers et al., 2011), and the build-up of pest resistance has led farmers to increase the frequency and intensity of pesticide applications over time.

To collect socio-economic and agricultural production data, a structured questionnaire survey was carried out in the Mae Sa watershed, which is comprised of twelve villages that practice agriculture. Figure 3 shows the location of the study area within the region, as well as the location of individual villages in the watershed. 20% of the farm households in each of these villages were randomly selected, which resulted in a total of 295 farm households being used for the survey. A one-year recall period, from April 2009 to March 2010, was used for the face-to-face interviews, with information about the farm households themselves as well as the land-use and cultivation practices gathered. For each plot and each crop, respondents were asked about inputs, outputs, their pest problems and how they have tried to control them. If using pesticides, respondents were asked to give the common names of each, the number of times they sprayed them, the quantity of undiluted chemicals used, and the price and volume per container. For each pesticide mentioned, data were collected on the active ingredients they contained from traders, shops and producers. The resulting data set is hence rather unique, in that it provides detailed farm-level information on the quantities of active ingredients applied, as well as comprehensive information on other production factors and farm characteristics.

Figure 3: Location of the study area

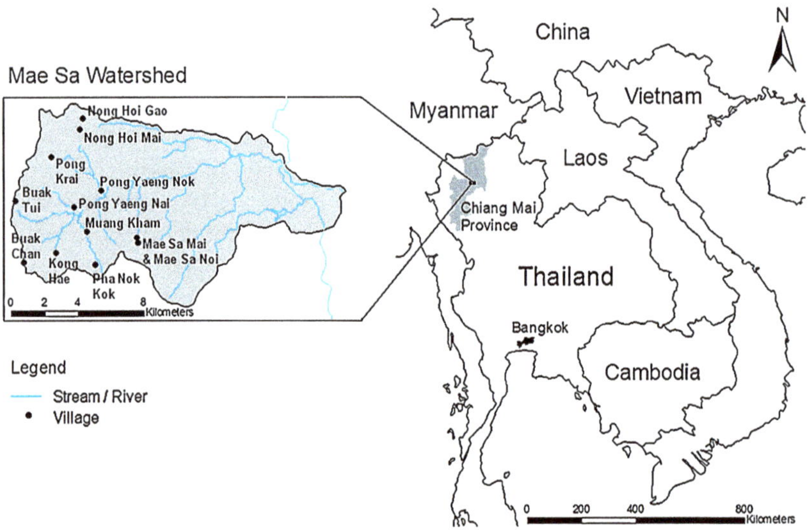

2.2 Farm characteristics in the study area

Compared to upland farmers in northern Thailand in general, farm households in the Mae Sa watershed are relatively wealthy, as for many years they have had resources, such as sprinkler and drip irrigation, greenhouses, Royal Project extension services and motorbikes, at their disposal. Due to the vicinity of the regional capital Chiang Mai, which guarantees good access to input and output markets, the watershed has commercialised earlier than other upland areas. Widespread intensive agriculture has been practiced since the early 1990s, with the process of intensification continuing to this day. Improvements in infrastructure have allowed farmers to reach a variety of product outlets, and to be able to sell them at better prices, as demonstrated by Schippmann and Quaim (2010) in the case of bell pepper production in the area. Table 3 shows that ethnicity clearly divides the population within the watershed. While the northern Thai farms are mostly located at mid-elevations, the villages inhabited by farmers of the Hmong minority are situated at higher elevations, and are generally less accessible. The household sizes in these Hmong villages are also much bigger, leading to a greater availability of on-farm labour. Liquidity per capita is lower among the Hmong households; however, with 28,000 Baht[1] as compared to 59,000 Baht for the northern Thai households, debt per capita is also lower. Farmers in the Hmong villages are generally of a lower formal education level, education being defined here as education higher than primary school.

Table 4 illustrates that land holdings in the study area are on average quite small, ranging from 0.71 ha in the northern Thai village of Pong Yang Nai, to 2.52 ha in the Hmong village of Mae Sa Mai. The higher population densities in the central watershed, where most northern Thai farmers live, leads to smaller farm sizes and more intensive production, with a greater use of greenhouses and the cultivation of flowers, as shown in Table 4. Buak Tui, where farmers specialize in growing roses, is an exception though, as it has a low proportion of people with Royal Project membership, and also has no farms certified with the public GAP standard. In the other Hmong villages, many farmers are members of the Royal Project, and use of the GAP certification scheme is more widespread than among the northern Thai farmers. The majority of Hmong farmers need to grow their crops on steep slopes, as they lack alternative locations among the higher elevations to be found in the watershed, and in several Hmong villages the litchi orchards represent a large share of the farmland used, at around 1 ha per household on average. The Royal Project used to promote litchi as a cash crop among Hmong farmers, in order to prevent soil degradation and as a replacement for opium cultivation.

1 Average exchange rate for 2013: 1 Euro = 40.86 Thai Baht (European Central Bank)

Table 3: Farm household (hh) characteristics and assets in the Mae Sa watershed villages, 2010

Village	Ethnicity	Hh size (Pers.)	Age of hh head (Years)	Educa-tion (%)	Liquidity per cap. (1000 baht)	Debt per cap. (1000 baht)	Own motor-bike (%)	On-farm labour (mandays/month/hh)	Off-farm labour (mandays/month/hh)	Hired labour (mandays/month/hh)
Pong Yang Nai	N. Thai	3.67	50.28	93	37.98	17.83	77	48.6	23.25	5.74
Pong Yang Nok	N. Thai	3.67	52.28	95	87.63	35.19	69	49.05	25.39	11.85
Pong Krai	N. Thai	3.82	45.29	94	64.66	35.82	82	59.59	20.71	6.86
Muang Kam	N. Thai	3.23	47.86	100	74.86	41.63	83	50.09	15.63	10.67
Kong Hae	N. Thai	3.76	50.62	100	29.12	20.99	67	40.57	21.81	8.08
Mae Sa Mai	Hmong	6.88	47.54	58	22.67	15.32	88	85.62	22.42	10.49
Mae Sa Noi	Hmong	6.18	45.12	59	14.24	3.58	82	60.82	29.24	11.41
Buak Chan	Hmong	6.86	48.86	57	48.93	13.58	81	98.29	15.33	20.31
Pha Nok Kok	Hmong	5.32	42.16	79	19.48	5.04	95	72.11	16.42	7.77
Buak Tui	Hmong	6.14	43.24	62	35.37	7.16	86	75.90	11.57	19.07
Nong Hoi Gao	Hmong	7.00	43.18	47	20.95	4.22	94	97.12	30.00	14.97
Nong Hoi Mai	Hmong	7.14	38.09	86	35.26	5.65	73	91.68	6.36	20.29

Table 4: Structure of farms in the Mae Sa watershed villages

Village	Farm size (ha)	Farm age (years)	Irrigated area (ha)	Litchi area (ha)	Flower area (ha)	No. of greenh. (#)	More than 1 crop (%)	Land w/o title (%)	GAP certific. (%)	Royal Project (%)
Pong Yang Nai	0.71	18	0.42	0.01	0.15	6.55	52	35	8	10
Pong Yang Nok	0.89	26.87	0.38	0.00	0.33	4.59	67	26	13	5
Pong Krai	1.15	19.06	0.23	0.00	0.29	11.47	35	51	0	6
Muang Kam	0.65	23.8	0.52	0.00	0.14	10.98	69	29	23	14
Kong Hae	0.86	25.86	0.71	0.04	0.00	1.33	86	55	10	52
Mae Sa Mai	2.52	27.42	2.07	1.32	0.00	2.62	77	100	42	69
Mae Sa Noi	1.75	24.57	0.86	0.82	0.00	1.41	71	100	41	35
Buak Chan	1.67	22.48	0.83	0.33	0.00	1.43	86	90	52	71
Pha Nok Kok	2.22	22.00	1.00	0.96	0.01	1.26	63	98	16	47
Buak Tui	2.10	23.67	0.24	0.31	0.50	1.62	62	100	0	33
Nong Hoi Gao	2.12	22.94	1.19	0.06	0.00	0.35	100	100	24	59
Nong Hoi Mai	2.22	18.5	1.07	0.03	0.00	0.63	100	91	27	68

2.3 Land-use in the study area

2.3.1 Description of cropping patterns

As mentioned above, the Mae Sa watershed is characterized by a very high diversity of cropping activities, which is a common feature of the upland areas in northern Thailand used for horticulture. The altitude gradients create numerous micro-climates which are suitable for different crop types. New crops have been introduced by traders (e.g. tomatoes) or by extension organizations such as the Royal Project (e.g. litchi), but not all villages in the watershed are equally accessible, so traders are more likely to deal with farmers closer to paved roads. The Royal Project, on the other hand, aims to promote the cultivation of cash crops among hill tribe farmers. As a result, many Hmong farmers nowadays grow litchi trees, the planting of which was supported technically and financially in the 1990s in the wake of large-scale efforts to find crops to substitute for opium cultivation. However, litchi trees are now increasingly being cut down and replaced by more profitable leafy, greenhouse and other vegetables (Schreinemachers et al., 2009; Schreinemachers et al., 2010). As prices fluctuate, farmers change their land-use accordingly. Nowadays, the main crops grown in the area are cabbages, kale, lettuce, bell peppers, tomatoes, chayote (*Sechium edule*), green beans, onions, chrysanthemums, roses and litchis, as well as rice (rainfed upland rice) and maize.

Pesticide use, as well as production costs and output, vary greatly among the various agricultural land-uses. Greenhouse vegetables, such as bell peppers and tomatoes for example, generate high sales revenues, but also entail high costs and require substantial expenditures on pesticides. As can be seen from Figure 4, farmers spend generally more on pesticides and other variable inputs for crops that generate relatively higher revenues.

Figure 6 meanwhile, illustrates the relationship between profitability and input intensity, showing the gross margins generated per hectare and per month together with pesticide use in kg per ha and per month. The need to protect valuable crops from virulent pests results in preventive as well as curative pesticide applications, which are extremely high for greenhouse vegetables, onions and flowers in particular.

Table 5 shows the average land area cultivated per farm for all the most important crops grown in the study area. The area cultivated is generally large for litchi, medium for flowers and small for vegetables. The table further shows sales revenues, pesticide use levels and costs, as well as the costs of other variable inputs.

Sales revenues per hectare and per month are highest for tomatoes, reaching 342,190 Baht, which for each cropping cycle amounts to 221,621 Baht based on

Figure 4: Crop outputs, total costs and pesticide costs for 15 crops in the Mae Sa watershed area, 2010 (in log Baht/ha/month)

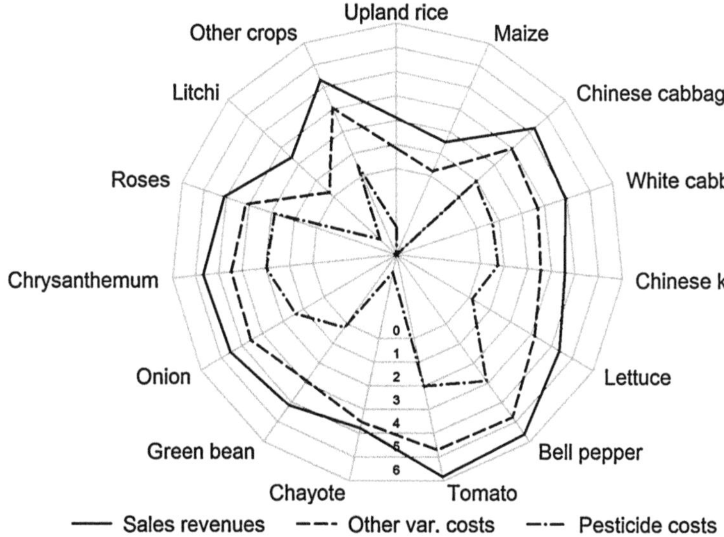

Figure 5: Gross margins and pesticide use for different crops (growing period in brackets)

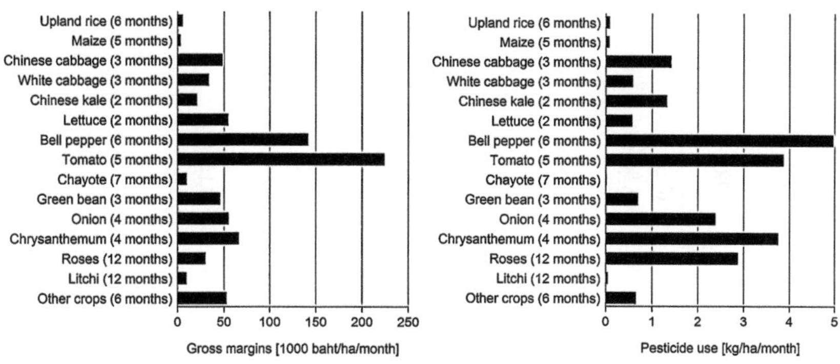

an average area of 0.12 ha for those farmers growing the crop, and an average growing period of 5.42 months. The total variable cost for a tomato cropping cycle is 76,330 Baht, resulting in a gross margin of 145,294 Baht (3,632 Euros) for an area of 0.12 ha over 5.42 months. Considering that the gross national income per capita in Thailand is 142,000 Baht (World Bank, 2013), this gross margin

Table 5: Average production data for important crops (standard deviations in brackets)

Cropping Activity	Farmers growing (%)	Area per hh (ha)	Growing length (months)	Labour (mandays / ha/month)	Pesticide use (kg/ha/ month)	Pesticide costs (1000 baht / ha/month)	Other var. Costs (1000 baht / ha/month)	Sales revenues (1000 baht / ha/month)
Upland rice	8	0.17	5.63	28.45	0.08	0.09	2.28	8.07
		(0.14)	(0.49)	(24.21)	(0.15)	(0.15)	(1.57)	(8.11)
Maize	13	0.17	4.63	29.46	0.08	0.03	1.24	4.57
		(0.16)	(0.49)	(67.85)	(0.38)	(0.09)	(1.11)	(4.22)
Chinese cabbage	24	0.17	2.50	105.24	1.41	2.54	19.56	70.17
		(0.14)	(0.50)	(77.59)	(1.78)	(2.45)	(10.75)	(66.94)
White cabbage	20	0.24	3.25	76.54	0.59	2.01	14.85	50.33
		(0.25)	(0.44)	(52.80)	(0.68)	(2.27)	(7.75)	(43.53)
Chinese kale	7	0.06	2.38	232.46	1.32	2.12	12.96	35.98
		(0.08)	(0.49)	(241.11)	(1.75)	(2.83)	(9.70)	(31.33)
Lettuce	11	0.11	2.34	139.79	0.57	1.15	23.88	79.48
		(0.10)	(0.48)	(108.39)	(0.89)	(1.55)	(13.53)	(67.05)
Bell peppers	32	0.13	5.72	548.66	4.97	18.16	117.61	277.68
		(0.12)	(1.21)	(4517.00)	(5.93)	(19.17)	(62.51)	(278.63)
Tomatoes	8	0.12	5.42	322.41	3.87	7.44	110.41	342.19
		(0.09)	(1.36)	(326.64)	(3.87)	(6.81)	(84.46)	(320.45)
Chayote	20	0.16	6.71	191.44	0.02	0.06	34.47	44.27
		(0.17)	(1.30)	(125.44)	(0.03)	(0.11)	(24.48)	(41.33)
Green beans	12	0.06	3.00	255.73	0.70	1.20	18.26	65.05
		(0.06)	(0.00)	(530.30)	(0.86)	(1.41)	(13.20)	(81.50)
Onions	7	0.10	4.00	151.94	2.39	3.99	34.91	93.94

Cropping Activity	Farmers growing (%)	Area per hh (ha)	Growing length (months)	Labour (mandays / ha/month)	Pesticide use (kg/ha/ month)	Pesticide costs (1000 baht / ha/month)	Other var. Costs (1000 baht / ha/month)	Sales revenues (1000 baht / ha/month)
		(0.06)	(0.00)	(96.68)	(2.83)	(5.34)	(15.68)	(78.17)
Chrysanthemums	17	0.25	4.40	204.11	3.75	7.21	32.06	105.81
		(0.27)	(0.49)	(226.57)	(4.14)	(8.28)	(19.24)	(103.89)
Roses	7	0.57	12.00	121.93	2.88	6.49	23.14	59.77
		(0.33)	(0.00)	(78.29)	(3.31)	(7.90)	(12.03)	(63.32)
Litchis	28	0.97	12.00	11.70	0.05	0.07	1.29	11.17
		(0.82)	(0.00)	(20.58)	(0.08)	(0.12)	(1.15)	(9.48)
Other crops	47	0.28	5.72	112.94	0.66	1.56	21.42	75.77
		(0.32)	(3.71)	(143.31)	(1.04)	(2.53)	(16.37)	(89.35)

is substantial, especially since two cropping cycles are possible in a year. So far, tomatoes have been grown by a few farmers only. Bell peppers however, are more widely grown, by 31.5% of farmers in the study area. It is also a high-value crop, generating as it does a gross margin of 106,912 Baht per cropping cycle over 5.72 months on an average 0.13 ha of land. The risk of losing bell-peppers, tomatoes and other valuable crops during pest attacks is considerable, and so farmers spray excessively. As witnessed by staff of the Uplands Program during many field visits, farmers frequently complain that the virulence of pest attacks has increased in recent years and that certain pesticides are no longer as effective or require large applications to produce the desired effect. There are clear signs of growing pest resistance and resurgence problems. Owing to the sizeable investment costs required, especially for greenhouses, and the high variable costs, many farmers are indebted (see Table 3) and so afraid of major pest damage events.

2.3.2 Categorisation and selection of land-uses

In total, 58 crops were recorded as part of the survey, despite the fact that over the years and across all the farms, more crops are likely to have been cultivated. However, the majority of these are minor crops, which are, when looking at the farming systems used as a whole, not very significant in terms of harvested area, pesticide applications and sales revenues. Figure 6 allow us to evaluate the relative importance of different land-uses, with the land-use categories used in the figure produced by grouping different crops together.

Figure 6: Cultivated area, pesticide use and sales revenues according to different land-use groupings

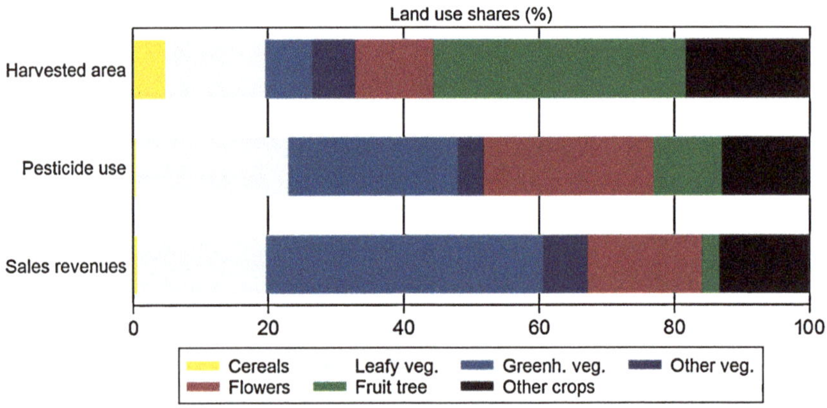

For the purpose of assessing pesticide use reduction strategies, this study focuses here on the most important crops economically, which tend to be those sprayed most intensively, as well as on those crops which cover large areas of the watershed. The following crops belong to the first category: Chinese cabbage, white cabbage, Chinese kale and lettuce (leafy vegetables), bell peppers and tomatoes (greenhouse vegetables), chayote, green beans and onions (other vegetables), and chrysanthemums and roses (flowers). The crops covering a large land area include upland rice and maize (cereals) and litchis (fruit trees). As there were too few observations to reliably estimate gross margins, various minor crops, denoted as 'Other crops' in Figure 6, had to be excluded from any further analysis and modelling. Also, for each of the three variables presented in the figure, the share of other crops is less than 20%, so omitting these crops does not constitute an oversimplification, and is justified. The diversity of crops is, of course, important for agro-biodiversity and potentially for breaking pest cycles, but very few farmers use these crops for the purpose of controlling pests.

Furthermore, the great diversity of cropping activities proved a challenge for the empirical analysis of pesticide productivity. Production functions are ideally estimated separately by crop, as different crops respond differently to fertilisers and other inputs, and are affected by pests in varying ways. However, the use of aggregate production functions, estimated at the farm, regional or even the country level, is also common in the literature (Carrasco-Tauber and Moffitt, 1992; Mundlak et al., 1997). As there were not enough observations to estimate separate production functions for each crop, the aggregated land-use groups listed above were used instead. Production functions were estimated for some of these groups based on similarities in terms of length of growing period, the pest problems experienced and the pest management activities used. In this context, greenhouse vegetables (including bell peppers and tomatoes) and leafy vegetables (cabbages, kale and lettuce) were the two most consistent groups, which is why they were used for the more detailed analysis of pesticide overuse.

2.4 Pest pressure, pest management and pesticide use in the study area

The intensification of agriculture has been accompanied by heightened pest pressure and the development of pest resistance for some crops. For example, farmers growing bell peppers, which as explained above are one of the most profitable crops in the area, struggle to control thrips, viruses and powdery mildew. With litchis fruit borer, shield bugs and downy mildew are major pests. Added to this, cabbages are frequently infested by webworms, beet army worms, common cutworms, cabbage loopers and diamondback moths. The graph on the left in Figure 7 gives an indication of the substantial pest pressure experienced in the study area, expressed as yield losses due to pests and diseases for the different land-use groups.

As a result of the significant threat posed by a multitude of pests, farmers try to protect their market crops by resorting to a vast array of chemical fungicides and insecticides. Schreinemachers et al. (2011) estimated that farmers in the watershed use an average of 13 kg/ha of active ingredients per year, which is high when compared with the average application rate of about 3.6 kg/ha per year for Thailand as a whole (Praneetvatakul *et al.* 2011). The right graph in Figure 7 demonstrates that the application of chemical pesticides is the predominant pest management strategy used in the study area. Farmers in the watershed depend very much on synthetic substances for pest control purposes, with non-synthetic methods, such as bio-pesticides, traps or mechanical control being practiced only in a small number of cases, based on the cropping observations made in this study. However, based on these same observations, in 18% of cases, no pest control takes place at all, either due to the low market value of the crops involved (e.g. litchis), because the crop is consumed on the farm (e.g. rice), or due to a lack of any severe pest problems (e.g. chayote).

Figure 7: Pest pressure in terms of yield losses and the use of different pest management methods

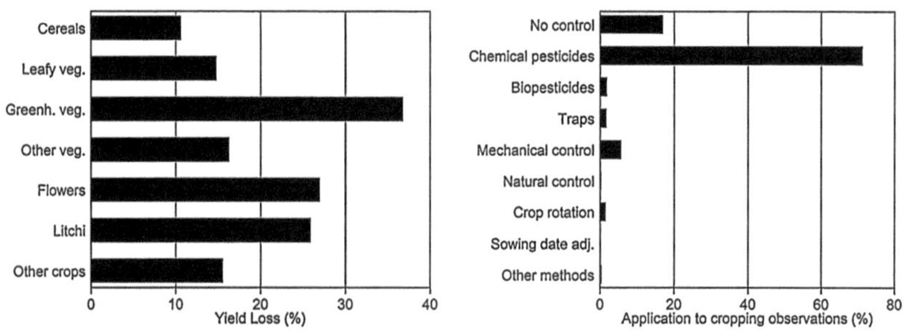

In Section 2.3.1 Figure 5 already gave a graphical overview of pesticide applications in the study area by crop. The levels of use are particularly high for tomatoes, bell peppers, flowers and onions, and are also elevated for cabbages, kale and lettuce. In their study, Schreinemachers et al. (2011) observed that farmers growing bell peppers used on average three times more pesticides than Spanish farmers, and 52 times more than Dutch farmers to produce the same quantity of output.

In the study area, applications are not homogenous across farmers. While half of the leafy and greenhouse vegetable growers in the study area use less than 1 kg/ha/month of pesticides, one fifth of farmers vastly exceed recommended input levels applying more than five and up to 20 kg/ha/month. There are also other strong indicators of heavy pesticide use in the study area. For instance, more than half of all farmers in the study area experience serious health problems after spraying, and three-quarters are concerned about their high levels of expenditure on pesticides. Furthermore, rivers in the area are heavily contaminated with pesticide residues, especially during the rainy season (Sangchan et al., 2012).

For most land uses, the majority of pesticide applications involve insecticides and fungicides, apart from cereals and fruit trees, where high quantities of herbicides are applied (see Figure 8). The main insecticides used are *abamectin* and *cypermethrin*, while *mancozeb* is the most commonly used fungicide. Farmers also resort to toxic substances such as *mevinphos* on a regular basis, as these quickly eliminate pests. The WHO toxicity classification sheds light on the hazardousness of different pesticides (WHO, 2009), giving an indication of the risks they pose to human health, which of course also reflects the risks posed to other living organisms. It is the most widely used classification of pesticide toxicity, and enables researchers and policy makers to quickly differentiate the more harmful from the less harmful substances. Pesticides ranked as WHO 1a and 1b are extremely hazardous, those ranked as WHO 2 are considered moderately toxic, while the WHO 3 toxicity class refers to slightly toxic pesticides. WHO U pesticides are described as unlikely to cause any harm. As Figure 8 shows, the proportion of applied pesticides belonging to different toxicity classes varies among the land-use groups. Not only are the largest quantities of pesticides applied on greenhouse vegetables and flowers, but also the most hazardous substances, while the proportion of moderately hazardous pesticides used is largest for leafy vegetables.

2.5 Vegetable IPM, the Royal Project and sustainable intensification

FAO defines IPM as "the careful consideration of all available pest control techniques and subsequent integration of appropriate measures that discourage the development of pest populations and keep pesticides and other interventions to levels that are economically justified and reduce or minimize risks to human health and the environment. IPM emphasizes the growth of a healthy crop with the least possible disruption to agro-ecosystems and encourages natural pest control mechanisms" (FAO, 2013b). In this sense, IPM can be described as an ecosystem approach used to manage pests and crops, incorporating factors such

Figure 8: Proportion of pesticides used by different types and WHO toxicity classes

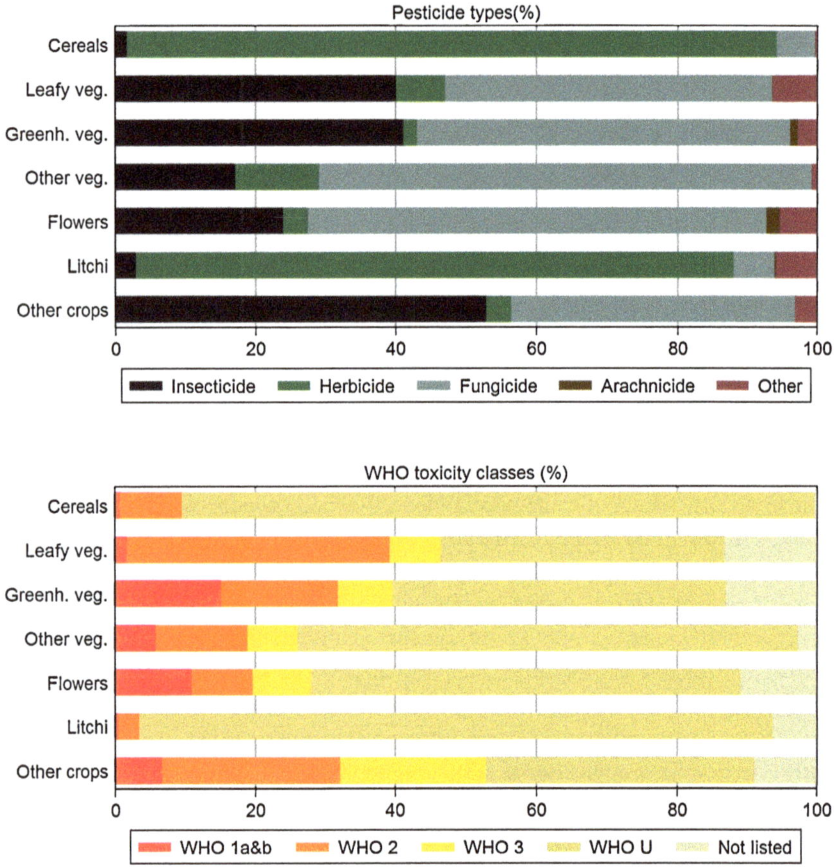

as the preservation of healthy soils, use of a diversity of cropping patterns and the conservation of beneficial insects. To better control pests, farmers are required to develop knowledge of the agro-ecosystem in question, and need to regularly observe their fields. Cultural, biological, genetic, mechanical and, as a last resort, chemical methods can be combined in a way that guarantees the long term environmental and economic sustainability of the farm, and that avoid harm to farmers' and consumers' health. In reality, IPM covers a vast spectrum of activities, and a pest control system is already referred to as integrated pest management, even if only a few of the above mentioned aspects are observed. This study uses the concept of IPM in a stricter sense.

While generally the practice of IPM among vegetable growers in the uplands of northern Thailand has remained rather insignificant, one pilot project of the national IPM programme, which has implemented integrated pest control, is a good example of how the sustainable intensification of vegetable production can work. Farmers belonging to the Royal Project station in Doi Angkhang have, with the support of the station's extension workers, successfully practiced IPM for several years now by following the above mentioned ecosystem approach to crop production and protection, and by combining different management strategies and farming practices in order to minimize the use of pesticides and grow more healthy vegetables. Growers of leafy vegetables at Doi Angkhang combine cultural (rotations and vigorous soils), biological (high agro-biodiversity with many natural predators) and mechanical methods (traps) with well-monitored bio-pesticide applications.

The Royal Project was founded in 1969 at the initiative of the King of Thailand to advance the sustainable development of Thailand's highland regions (Sathaban Wichai et al., 2007). Initially, the principal aim of the Project was to replace the cultivation of poppy for opium production by promoting alternative crops among hill tribe farmers. Poverty among upland communities, soil erosion and deforestation were other problems addressed by the Royal Project and which gradually came to the fore with the success of the opium eradication programme. A network of research and extension stations was set up throughout northern Thailand to test and develop suitable and economically profitable crops, to advise farmers locally, and to support them with the marketing and distribution of their products. Among these stations, the Royal Project operation in Doi Angkhang was one of the first to be set up, and is one of the largest facilities in the Thai highlands. It is also provides a model for farming activities in mountainous subtropical areas, and acts as a research and rural extension service at the same time. As such, it directly serves the surrounding village communities, where farmers from a wide range of ethnic backgrounds live, and aims to improve the livelihoods of these farmers through the use of sustainable intensification processes. Besides leafy vegetables, the cultivation of tea and coffee, berries, fruit trees and flowers – all of which are crops that need a more temperate climate – is promoted by the station.

Table 6 shows data for three typical rotations of the three of the main leafy vegetables grown by the farmers in Doi Angkhang: cabbages, lettuce and spinach. Data was collected following these rotations in 2012, representing one-third of all those using IPM in Doi Angkhang. The other two-thirds were practicing distinct vegetable rotation systems that year, for which no data was collected. Since production is very homogenous among farmers and strictly controlled by the Royal Project staff, a relatively low number of representative cropping observations was sufficient.

Table 6: Production data for IPM vegetable rotations as practiced by farmers at Doi Angkhang (2012, n = 34)

Season	Crop	Growing length (months)	Labour requirement (mandays / ha/ month)	Bacillus turingh. (kg/ha/ month)	Bacillus subtilis (kg/ha/ month)	Tricho-derma (kg/ha/ month)	Azadi-rachtin (kg/ha/ month)	Metazan (kg/ha/ month)	Pesticide costs (baht/ha/ month)	Other var. costs (baht/ha/ month)	Sales revenues (baht/ha/ month)
Rotation option 1: Cool season cabbage -> Hot season lettuce -> Rainy season spinach											
Cool	Cabbage	3	382	22.75	3.97	3.97	1.85	1.39	12,313	15,848	64,815
Hot	Lettuce	3	268	3.17	2.38	2.38	0	0	2,056	11,954	79,365
Rainy	Spinach	1	1,430	7.54	5.66	5.66	0.60	2.08	7,632	27,433	463,316
Rotation option 2: Cool season lettuce -> Hot season spinach -> Rainy season cabbage											
Cool	Lettuce	3	378	5.36	22.75	3.97	3.96	1.85	12,313	20,319	164,931
Hot	Spinach	1	589	11.65	3.17	2.38	2.38	0	2,056	24,739	133,890
Rainy	Cabbage	3	255	2.38	7.54	5.66	5.65	0.59	7,632	18,130	44,643
Rotation option 3: Cool season spinach -> Hot season lettuce -> Rainy season cabbage											
Cool	Spinach	1	451	9.00	5.36	4.32	4.32	0	4,547	17,851	127,538
Hot	Lettuce	3	268	3.17	11.64	5.95	5.95	0	7,130	11,954	79,365
Rainy	Cabbage	3	284	19.52	2.38	5.36	5.36	0	4,760	13,800	41,984

At 0.16 ha, the area of land per farmer is very limited in the area, and this is the main constraint on being able to increase output mentioned by the interviewees. Improving the livelihoods of farmers therefore involved intensifying production while minimizing environmental impacts on the unique mountain ecosystem. The usual size of one plot is only 60 square meters, and around half of this is covered with vetiver grass to prevent erosion. The vetiver grass strips are criss-crossed with many other species of flowers, herbs and grasses, and harbour a great variety of beneficial organisms. The surrounding forests, as well as the fallow plots, allow for a large diversity of natural predators to exist. After each crop, farmers keep their plots fallow for several weeks, often planting beans, which improve the soil quality and break the pest cycle. Each crop is managed according to a recommended cultivation plan developed by the Royal Project station. For cabbages, this involves the application of 3 to 5 kg of manure after planting, the spraying of a diluted organic fertilizer which has been produced from vegetable scraps, molasses and microorganisms every three days, and the application of bio-pesticides such as *Bacillus thuringensis* for use against worms, and *Bacillus subtilis* for use against parasites. *Trichoderma, azadirachtin* and *metazoan* are other bio-pesticides at the disposal of the IPM farmers, for which pre-specified quantities are applied at regular intervals, depending on the season. Specific, substances or higher amounts are used when pest pressure crosses a particular threshold. Farmers closely observe their plots, and also resort to traps and hand-picking to protect them.

For each cropping cycle and plot, the Royal Project obliges farmers to keep detailed records, and subsequently monitors applications, to make sure the recommended amounts are not exceeded. To guarantee the safety of its produce, staff from the Royal Project continuously test vegetable samples for residues.

The ex-ante assessment of pesticide use reduction strategies using an agent-based model allows one to include sustainability innovations as an alternative to current cropping activities. The data presented in Table 6 attempts to simulate the diffusion of IPM methods for leafy vegetables in the MPMAS model, which was otherwise built with data collected from the Mae Sa watershed only. The lack of data available on alternatives to chemical pest management from the survey conducted in the watershed, necessitated that observations from a similar environment be used. The climate and terrain of the Royal Project station at Doi Angkhang, which is located at 1400 m above sea level, is similar to that of the central and particularly the upper Mae Sa watershed. The land-use mix is also comparable, as leafy vegetables are an important crop in both places, and with the pest complex being very much alike. Average yields across the three crops are 25% lower for farmers at Doi Angkhang, but prices are higher and more uniform, and lie within the upper tercile range for leafy vegetable prices in the Mae Sa watershed. It should also be noted that variable input costs for IPM are lower, while labour requirements are higher. A substantial difference between farmers in both locations can be seen in the marketing of the produce. Farmers in Doi Angkhang have a secure retail buyer in the form of the Royal Project, which provides full logistics facilities in terms of transport and sales. The vegetables from Doi Angkhang can be also be sold at a mark-up due to certification with the Organic Thailand label. Farmers in the Mae Sa watershed have a variety of buyers for their produce, including the Royal Project among other outlets such as middlemen or direct sales at the market. Since the Royal Project's infrastructure is well established in the watershed, as well as in many other parts of northern Thailand, it is possible to introduce an innovation brought-in at one station, at another. Increased awareness of food safety issues among Thai consumers in recent years has translated into a greater willingness to pay for healthy vegetables, and justifies keeping the prices for the IPM produce higher.

3. Methods

3.1 Quantification of pesticide productivity and pesticide overuse from farmer as well as from societal points of view[2]

3.1.1 Conceptual frame

In line with previous studies, pesticide overuse here is defined as the amount of pesticides used in excess of an economically-defined optimum (Huang et al., 2002; Jah and Regmi, 2009; Qaim and De Janvry, 2005; Sexton et al., 2007). Making the simplifying assumption that farmers are motivated to maximize their profits, a private optimum level of pesticide use can mathematically be derived as being the point at which the marginal returns associated with pesticide use equal the farmers' marginal purchase costs for those same pesticides (i.e. the purchase price). A social economic optimum includes the negative externalities of pesticide use, being the point at which the marginal returns are equal to the sum of the marginal purchase cost and the marginal external cost.

The marginal returns for pesticide use can be derived from a production function analysis. Lichtenberg and Zilberman (1986) argued that treating pesticides in the production function as a damage-control agent rather than a regular growth-stimulating input avoids overestimating the efficiency of pesticide use, a phenomenon confirmed by successive studies (Chambers and Lichtenberg, 1994; Praneetvatakul et al., 2003; Shankar and Thirtle, 2005). Following Lichtenberg and Zilberman (1986), for this study crop output (Y) is thus specified as a function of growth-stimulating inputs $F(Z)$ and damage control agents $G(X)$:

$$Y=F(Z)G(X) \tag{1}$$

The function $G(X)$, which has a value of between zero and one, thus determines the magnitude of any damage and the effectiveness of any control applied using pesticides (X). In accordance with the original framework proposed by Lichtenberg and Zilberman (1986) and most of the related economic analyses involving pest damage, separability between potential output and losses is assumed. This assumption implies that damage does not depend on potential output, i.e. the effectiveness of damage control is independent of the mixture of direct inputs,

2 This section has already been published as part of the following article: Grovermann, C., Schreinemachers, P., Berger, T., 2013. Quantifying pesticide overuse from farmer and societal points of view: An application to Thailand. Crop Protection 53, 161–168.

and that $F(Z)$ exhibits constant returns to scale (Carpentier and Weaver, 1997; Kuosmanen et al., 2006). By introducing prices for output (p) and inputs (w for growth-stimulating inputs, v for pesticides), the farm-level profit function is specified as:

$$\Pi = pY - wZ - vX \qquad (2)$$

Maximizing this function with respect to pesticides gives us the private economic optimum level of pesticide use:

$$d\Pi/dX = 0 \quad \text{or} \quad d(pF(Z)G(X) - wZ - vX)/dX = 0 \qquad (3)$$

Not all costs associated with pesticides are transmitted through the price (v) that farmers pay for them. Being toxic by design, pesticides can harm organisms other than pests, such as beneficial insects and soil organisms, aquatic life and humans. Costs to society are incurred in the form of pest resurgence and pesticide resistance, chronic and acute health problems experienced among applicators or pickers, and for consumers through ingesting pesticide residues and due to the contamination of water sources. Costs are also incurred in the form of pesticide monitoring by governments. These costs are called external costs (Praneetvatakul, 2013; Pretty, 2000). Including them in the price of pesticides will raise their overall cost and lower the optimum level of pesticide use. This is illustrated in Figure 9.

Figure 9: The private and social optimum level of pesticide use

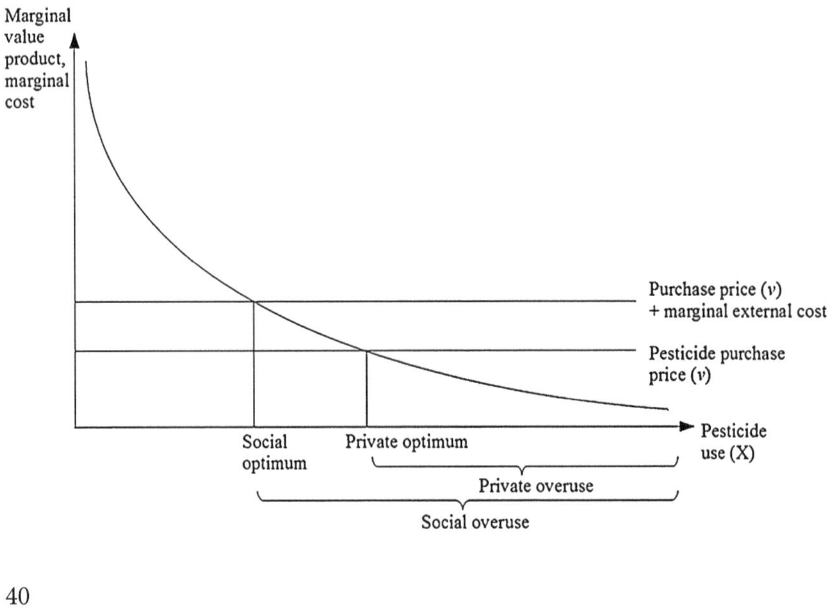

Graphically, the private optimum is represented by the intersection of the marginal benefit of pesticide use, which can be derived from the production function analysis and is referred to as marginal value product, and the marginal private cost of pesticide use, which corresponds to the pesticide purchase price. The social optimum is given by the intersection of the marginal value product and the marginal social cost, i.e. the external marginal cost added to the private marginal cost. The PEA tool, as explained in Section 2.4 below, allows quantifying the marginal external cost such that it can be added to the marginal private cost in order to determine optimal use from a societal point of view.

3.1.2 Specification of the production functions

The Cobb-Douglas and quadratic specifications are the most commonly used types of farm production functions, and have been shown to give similar results (Horna et al., 2008; Qaim and De Janvry, 2005). Unlike the Cobb-Douglas method, the quadratic form allows for decreasing total and negative marginal returns to be considered, and can also handle zero values for input or output variables, though multi-collinearity is a frequently encountered problem. The Cobb-Douglas function, on the other hand, tends to give better results if inputs and outputs have a high variation, as the logarithmic transformation reduces the spread in values. In this study, there were few zero values but a relatively high variation in observed values. For the set of growth stimulating inputs in *F(Z)*, it could be considered reasonable to assume diminishing but not negative marginal returns. The curve of the total and marginal value product flattens out at higher levels of input, which may be a disadvantage if the economic optimum occurs at such high levels. However, the results of this study suggest that this was not the case here. The Cobb-Douglas function further assumed constant returns to scale as well as convexity to the origin, implying some complementarity among inputs, but no full substitution. An F-test for restrictions was used to test the restricted Cobb-Douglas specification against a flexible translog specification, which nests the Cobb-Douglas function. Based on this test, the use of the Cobb-Douglas specification could not be rejected (the F statistic was 0.830 ($p = 0.363$) for leafy vegetables and 0.210 ($p = 0.649$) for greenhouse vegetables).

Various specifications have also been proposed for the damage abatement term *G(X)*, such as exponential, logistic, Pareto and Weibull (Lichtenberg and Zilberman, 1986). As several recent studies have shown, the exponential specification gives robust results (Jah and Regmi, 2009; Pemsl et al., 2005; Skevas et al., 2012) so this specification was employed in the present analysis. It is defined as follows:

$$lnY = \alpha + \sum_{i=1}^{n} \gamma_i C_i + \sum_{j=1}^{m} \beta_j \ln Z_j + \ln[1 - e^{-\lambda X}] + \in \quad (4)$$

The constant α and the coefficients γ_i, β_j and λ for this function were estimated for two distinct land uses in the study area, these being leafy vegetables and greenhouse vegetables, because these have very different output levels (Y) and use a different technology (open field vs. closed system). Within each land use management type, the growing periods and pest problems experienced are similar. The indicator variables C_i were introduced alongside growth-stimulating inputs Z_j and pesticides X, to control for farm characteristics. These farm characteristics also included crop and location dummies, those that captured differences in crop management and agro-ecological conditions.

As explained above, the private optimum of pesticide use occurs at the point, where the marginal value product equals the purchase price of pesticides. However, in this study, pesticides are expressed in monetary rather than physical quantities, which means that the purchase price is included in the pesticide variable. As a consequence, the optimal private level of pesticide use is found to occur where the marginal value product equals unity. Likewise, the social optimum is obtained where the marginal value product equals unity plus the ratio of external costs to pesticide purchasing costs.

Thereby the marginal value product of pesticides describes the change in the value of output that results from spending one more monetary unit on pesticides. It is the first derivative of the production function in equation (4), and is specified as:

$$MVP_x = F(Z)^* \lambda[\exp(-\lambda X)]/[(1-\exp(-\lambda X)] \quad (5)$$

Equation (5) shows that the marginal value product of pesticides is observation-specific, because it depends on the level at which all other inputs are applied. Therefore, the total quantity of pesticide overuse was computed for the study area by summing differences between the actual pesticide use (X_a) recorded in the survey, and the optimal pesticide use (X^*) calculated from the observation-specific marginal value product:

$$\text{Total Overuse} = \Sigma\,(Xa - X^*) \quad (6)$$

3.1.3 Econometric estimation of pesticide productivity

Parameters were estimated using non-linear least squares regressions with robust standard errors. The Variance Inflation Factor was found to be well below

10 for each regression, suggesting that multi-collinearity might not be a problem here. In addition, it was necessary to control whether pesticides were an endogenous variable in the model, as several previous studies have found it to be correlated with the error term. A test for endogeneity was therefore conducted using a two-stage least-squares (2SLS) instrumental variable regression following Horna et al. (2008) and Huang et al. (2002), which provided no evidence that pesticide use was endogenously determined (Wu-Hausman F value of 0.604 ($p = 0.438$) for leafy vegetables and of 2.293 ($p = 0.132$) for greenhouse vegetables).

For the regression analysis, pesticide amounts as well as crop output, fertilizer amounts and other inputs were all expressed in Baht per hectare per month. Just like the pesticide data, all output and other variable input data were recorded for each plot and each crop that farmers were growing. The variable 'Fertilizers' comprised data on the quantity of mineral and organic fertilizer valued by their price. The variable 'Other inputs' included seed or seedlings, plant hormones as well as planting material valued by the respective price.

Table 7: Summary statistics of variables used in the analysis

Variables	Leafy Vegetables		Greenhouse Vegetables	
	Mean	SD	Mean	SD
Spraying method (1=preventive)	0.59	0.49	0.51	0.50
Education (1=low)	0.46	0.50	0.48	0.51
Irrigation (1=using)	0.47	0.50	1.00	0.00
Location 1 (1=Thai villages at high altitude)	0.06	0.23	0.16	0.37
Location 2 (1=Hmong villages at high altitude)	0.09	0.29	0.20	0.40
Location 3 (1=Hmong villages at high altitude)	0.76	0.43	0.09	0.29
Output (1000 baht/ha/month)	46.05	39.80	213.47	187.21
Labour (hrs/ha/month)	95.77	96.32	246.88	219.22
Fertilizers (1000 baht/ha/month)	7.10	4.54	53.18	39.79
Other (1000 baht/ha/month)	1.77	1.26	36.72	24.55
Pesticides (1000 baht/ha/month)	1.71	1.67	15.16	14.82
External costs (1000 baht/ha/month)	0.49	0.59	3.21	3.56

Notes: Omitted location dummy is Thai villages at middle altitude. The crop dummies are not shown.

Other explanatory variables included crop dummies, village dummies, irrigation, education and the spraying habit of farmers. In the production function analysis for leafy vegetables, crop dummies included white cabbage, Chinese cabbage and kale, while for greenhouse vegetables a tomato dummy was used. Other variables control for farm and farmer characteristics, such as the ethnicity and location of the village related to a particular observation, because data were collected from structurally different Thai and Hmong villages at middle and high altitudes. As no data were available on the amount of irrigation water, a dummy was included indicating whether irrigation was used or not. As education levels differed among farmers, which might impact on production, the education dummy 'Low education' specified if farmers attended at most primary school. Likewise, the spaying habit of farmers differed. Here, data were recorded on the predominant pesticide application strategy of farmers, preventive versus curative. Table 1 summarizes for each group the variables used in the analysis.

3.1.4 Quantification of the external costs of pesticide use

As mentioned before, the data collected in the Mae Sa watershed is rather unique in that it provides detailed farm-level information on quantities of active ingredients, which are required for using the Pesticide Environmental Accounting (PEA) tool. With the help of this tool, external costs could be calculated for each individual active ingredient and then be aggregated to an overall external cost estimate.

The PEA tool was developed by Leach and Mumford (2008, 2011). It is a cost-transfer approach that was calibrated from detailed actual cost studies carried out in Germany, the UK and the USA (Pretty et al., 2000; Pretty et al., 2001). Comprehensive information was available in this study on the costs of monitoring pesticide use, of remedying damage to ecosystems and of treating pesticide-related health problems. These actual cost data were used as external cost base values in the analysis, and then 'transferred' to other countries by adjusting for different application rates, the toxicity of applied pesticides, as well as economic conditions.

The tool allocates the external costs of pesticides to particular pesticide compounds based on application rates and potential risk. For potential risk, it uses toxicological data on the harmful effects of pesticide compounds on applicators and pickers (farm workers), on groundwater leaching and in terms of the pesticide residues left on food (consumers), as well as on aquatic life, bees, birds and beneficial insects (the environment). These toxicological data were taken from the Environmental Impact Quotient (EIQ) tool developed by Kovach et al. (1992).[3]

---

3 EIQ base values are available from an online database: http://cceeiq-lamp.cit.cornell.edu/EIQCalc/input.php (accessed January 2011).

Using two economic adjustment factors, Leach and Mumford (2008) then 'transferred' these costs to estimate the external costs of pesticide use for agricultural production systems in Spain, Turkey and Israel. Praneetvatakul et al. (2013) tested the use of the PEA tool for Thailand by comparing the PEA estimates at the national level to an accounting of actual pesticide costs for two years. Although the PEA tool overestimated actual costs in one year and underestimated the costs in the other year, on average over all years the estimates were in a similar order of magnitude.

Based on the PEA method, the total external cost (TEC) of a pesticide *p* can be calculated as:

$$TEC_p = Rate_p * \frac{Active_p}{100} * \sum_{c=1}^{8} \left[EC_c * F_c * (F_{agemp} \,|\, c = 1,2) \right] * F_{gdppc} \tag{7}$$

$Rate_p$ is the application rate of a pesticide *p* in kg of formulated product per hectare, and $Active_p$ represents the percentage of active ingredient contained in the formulated product. The EIQ methodology uses eight categories (*c*=1,2,..,8) to distinguish the eco-toxicological effects of active ingredients. EC_c denotes the base value of the external cost attributed to category *c* (*c*=1,2,..,8). As pesticides with a higher potential risk should be associated with a higher external cost, the potential risk is divided into three categories. In doing so, lower, medium and upper values are multiplied with the external costs by a factor (F_c) of 0.5, 1.0 and 1.5, respectively. Leach and Mumford (2008) defined the three levels of factor (F_c) according to low, medium and high toxicity ranges based on the EIQ. F_{agemp} and F_{gdppc} are adjustment factors used to indicate the importance of employment in agriculture and the costs of pesticide monitoring and clean-up activities respectively, as further explained below.

The effects of pesticides on applicators and pickers (i.e. farm workers) are likely to be greater in low-income countries due to the fact that relatively more people are employed in agriculture and thus come into direct contact with pesticides. Whereas Leach and Mumford (2008) proposed the proportion of GDP taken up by agriculture as a proxy for health-related externalities, Praneetvatakul et al. (2013) preferred using the share of agriculture in employment terms. It is considered to better reflect the number of people who tend to be exposed to pesticides on farms. The external costs for applicators and pickers (*c*=1,2) are thus multiplied by a factor $F_{agemp,}$ calculated as the ratio of a country's share of employment in agriculture to the average share of agricultural employment in Germany, the UK and the USA (weighted by GDP). The authors point out that this approach does not capture the fact that pesticide use in low-income countries is far more hazardous, because of a lack of sufficient protection. On the other hand, as lower labour costs reduce expenditures of monitoring and clean-up, low-income countries are supposed to incur fewer external costs. Therefore Leach and Mumford (2008) resorted to the adjustment factor F_{gdppc}, which is calculated as the ratio of a country's per capita GDP to the average per capita GDP in Germany, the UK and the USA (weighted by GDP). Multiplying the total external costs with this factor thus allows taking into account the fact that labour is cheaper in a developing country.

Praneetvatakul et al. (2013) applied the PEA tool to the same site as in this study, using the same farm-level data. Based on an average application rate of 13 kg of active pesticide compounds per hectare in 2010, they estimated an average external cost of USD 106/ha, which compares to average pesticide expenditures of USD 963/ha. They estimated that internalizing the external costs into the price of pesticides would increase the price of pesticides by about 32% in the study area. For more details, refer to Praneetvatakul et al. (2013) as the methods and data are exactly the same.

Quantifying the external costs using the PEA tool is an essential part of analysing pesticide overuse from a societal point of view. To do this, external costs are required in order to obtain the social costs of pesticide use and then derive their marginal social costs. At the point where the latter are equal to the marginal value product, the optimal level of pesticide use can be determined. Pesticides were expressed in monetary units rather than physical quantities here, so the social optimum was in fact computed where the marginal value product equals unity plus the ratio of external costs to pesticide purchase price.

3.2 Innovation diffusion and adoption probabilities

3.2.1 Agricultural technologies and the theory of innovation diffusion

Innovations in agriculture range from rather simple technologies such as new fertilisers, improved varieties or the use of certain machinery, to more complex tools that involve fundamental cropping system changes. The introduction of integrated cultivation methods, which for example relies on soil conservation techniques and biological pest management, is very knowledge intensive and requires the fine-tuning of many production parameters. Regardless of the type of technology, it is assumed that the diffusion of innovations in the study area follows a similar pattern. According to Rogers (2003), the adoption process consists of five stages:

In the case of a new agricultural production technique, farmers only take the adoption of an innovative technology into account after gaining knowledge of it and being persuaded by a peer group to actually test it. These two early stages of

Figure 10: Adoption process according to Rogers (2003)

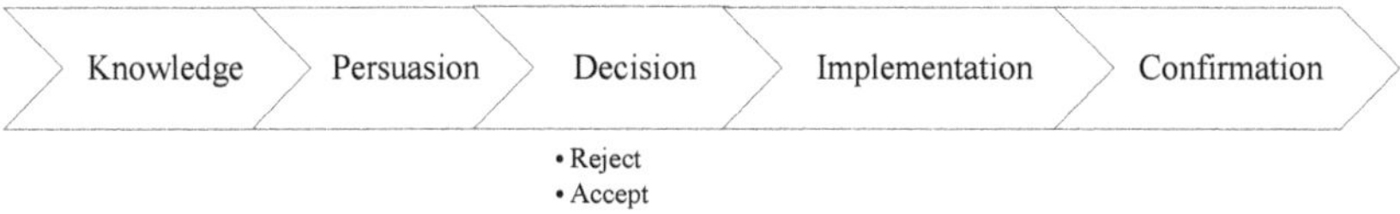

adoption, 'knowledge' and 'persuasion', involve various flows of information from external and internal sources, and intensive networking among farmers. In order to capture the role of peer-to-peer communications in the decision-making process on individual innovation adoption, Valente et al. (2003) suggested a so-called network-threshold model of innovation communication. According to this model, the more widespread the use of an innovation, the more effective is "social" persuasion by peers, and also farmers with rather high thresholds to adoption start considering the technology for themselves. This "frequency-dependant" or "bandwagon effect" is well documented within the sociology literature, e.g. Granovetter and Chung (1997).

The eventual individual decision to actually change cultivation practices, that is, whether to accept or reject an innovation, is made during the decision stage of the adoption process. Not every farmer has the same probability of accepting an innovation; it depends on the position of the individual in the network as well as his or her socio-economic characteristics. Rogers (2003) classifies individuals according to their innovativeness, and so divides them into five adopter categories: Innovators, Early Adopters, Early Majority, Late Majority and Laggards. Adding to this classification, and based on the concept of network communications, Berger (2001) postulates adoption to be contingent on the number of individuals who have already implemented the innovation. The frequency distribution underlying Rogers's adopter classification can then be interpreted as network thresholds. 'Early adopters' start considering a specific innovation only when the 'innovators' have already adopted it and 'exposure' to the innovation has reached the early adopter network threshold. Once the indicated percentage of individuals has adopted the innovation and the related threshold has been reached, the innovation spreads to the next adopter category. The network thresholds for the various adopter categories are 0%, 2.5%, 16%, 50% and 84% respectively.

The extent to which an innovation is eventually used is then determined during the implementation stage, and depends mainly on the innovation's usefulness to the farmer. In this context, implementation might happen fully, partially or not at all, and this degree of implementation is subject to factors such as individual resource constraints or the relative profitability of different production techniques.

3.2.2 Specification of the adoption regression model

The decision to accept or reject an innovation corresponds to the outcome of a particular course of action, which can be understood as a binary choice. The adoption ($y_j = 1$) vs. non-adoption ($y_j = 0$) choice can thus be used to predict adoption probabilities from a set of observable independent variables. Adoption here indicates whether a technology has ever been adopted, not whether it is used in the year of data collection. Since knowledge of the innovation is limited to a part of the population only, a two-stage econometric procedure is suggested here to better understand why some farmers adopt early and others wait a substantial amount of time before deciding to adopt a new practice. The first stage corresponds to the knowledge and persuasion parts of the adoption process, in which the awareness of the innovation among farm households is determined; the condition for these households to then consider adoption and for social network thresholds to be reached. Accordingly, the second stage involves an econometric estimation of the actual adoption decision. The first and second stage estimations are based on probit regression models. When the error terms of these two models are correlated ($\rho \neq 0$), standard probit techniques applied to the first equation yield biased results. A probit model with sample selection provides consistent, asymptotically efficient estimates for all parameters in such cases (Van de Ven and Van Praag, 1981). The actual dependent variable y^* constitutes the scale of adoption and is an unobservable magnitude. X_j includes a vector of strictly exogenous variables that determine adoption, while z_j includes a vector of variables that determine knowledge of the innovation. β_j and γ_j are the vectors of parameters to be estimated, and u_{ij} is the household specific error term. The model assumes that there exists an underlying relationship, as follows:

$$y_j^* = x_j \beta_j + u_{1_j} \quad \textit{latent equation} \tag{8}$$

Such that only the binary outcome is observed:

$$y_i^{probit} = (y_i^* > 0) \quad \textit{probit equation} \tag{9}$$

The dependent variable is however not always observed. Rather, the dependent variable for observation *j* is observed if:

$$y_i^{select} = (z_j \gamma_j + u_{2j} > 0) \quad \textit{selection equation} \tag{10}$$

And where:

$$u_1 \sim N(0; 1); \; u_2 \sim N(0; 1); \quad corr(u_1; u_2) = \rho \tag{11}$$

When $\rho \neq 0$, a probit model of adoption is required that corrects for selection bias. The probability of observing a positive outcome for adoption is given by the following equation:

$$\Pr(y|x)=\{\psi(\beta x_i)\}^{y_i}\{1-\psi(\beta x_i)\}^{1-y_i}, \quad yi = 0{,}1 \qquad (12)$$

A log-likelihood test of independent equations allows a check to take place as to whether the combined probit model with sample selection is superior to estimating using two independent models.

3.2.3 Innovativeness ranking and categorisation

Innovativeness is generally seen as a personal characteristic that distinguishes farm households, such that while the most innovative farmers eagerly test new technologies, ordinary farmers might be more reluctant to do so. Innovativeness is difficult to measure directly and so is not usually recorded in surveys of farm households (Schreinemachers et al., 2009). As a result, two key criteria are used to predict the innovativeness levels of farm households, based on information from the available survey data, as follows:

- Time lag between the year of introduction (technology promulgated for the first time in the watershed) and the individual adoption decision, and
- Econometrically estimated adoption probabilities (see section before).

The time lag between introduction and the individual adoption of an innovation is a strong indicator of how innovative a farm household is. Using adoption probabilities alone might neglect important information on diffusion, which is available from the survey data. Therefore, time lag information was used here to obtain an initial innovativeness ranking for the farm households. However this ranking could not be completed, since many observations were associated with the same time lag or an unknown time lag. These were, as a consequence, assigned the same rank. The time lag information was hence complemented by the predicted adoption probabilities, which made it possible to refine the ranking. Each farm household could be given a unique ranking in this way.

In a last step, the innovativeness ranking was transformed into an adopter categorization in line with the five adopter threshold groups. For this purpose, the threshold values of the adopter categories (0%, 2.5%, 16%, 50% and 84%) were matched with a cumulative distribution function of the innovativeness ranking. This allowed the determination, for each farm household, of whether they were an innovator, an early adopter, early majority, late majority or a laggard.

3.2.4 Econometric estimation of adoption probabilities

As explained above, farmers in the study area rely almost exclusively on chemical pesticides for their pest control. Recent government efforts have aimed to reduce these high levels of pesticide use by means of the voluntary public GAP standard.

In total, 20% of farmers in the Mae Sa watershed are GAP certified, at the national level certificates were issued to about 212,000 farmers covering a crop area of 225,000 hectares in 2010 (Schreinemachers et al., 2012). The GAP standard is as such the most widespread pesticide use reduction initiative, even though it has been found not to deliver profound pest management changes at the farm level. Nevertheless, it provides an entry point for analysing the determinants of innovativeness in this sector. In accordance with the two-stage econometric estimation procedure of adoption probability, the dependent variables are whether farmers know ($y_j^{select} = 1$) or do not know of ($y_j^{select} = 0$), as well as whether farmers do ($y_j^{probit} = 1$) or do not possess ($y_j^{probit} = 0$) GAP certification. The analysis is based on the explanatory variables listed in Table 8, which represent farm characteristics, such as land size and wealth for example, as well as network characteristics. Of the network characteristics, exposure measures the links of an individual farm household to those households which are aware of the innovation. The variable reflects the proportion of contacts that know about an innovation when compared to contacts that don't know, indicating levels of communication among farm households (Valente, 2005). Royal Project membership and owning a motorbike are proxies for the network attribute of cosmopoliteness, that is, the number of ties an actor has with the outside world. The variable "Village head" in turn acts as a measure of the importance or prominence of a person in a network, i.e. the centrality.

Table 8: Determinants of innovativeness

Variables	**Summary statistics**		**Correlation with time lag**	
	Mean	**St. dev.**	**Coef.**	**p-value**
Household size (#)	5.02	2.558	−0.138	0.018
Farm size (ha)	1.44	1.354	−0.127	0.030
Percentage of high value vegetables (%)	0.39	0.373	−0.114	0.050
Percentage of litchi (%)	0.13	0.252	−0.114	0.051
Farm age (years)	22.99	11.969	0.046	0.427
Age household head (years)	47.01	10.398	0.086	0.142
Education (yes=1)	0.81	0.395	−0.064	0.273
Grow more than 1 crop (yes=1)	0.71	0.455	−0.179	0.002
Applying pesticides regularly (yes=1)	0.40	0.491	−0.032	0.587
Liquidity per capita (1000 baht)	45.33	121.185	−0.026	0.655
Own motorbike (yes=1)	0.80	0.398	−0.144	0.013
Member of Royal Project (yes=1)	0.35	0.478	−0.472	0.000
Exposure (#)	0.42	0.250	−0.391	0.000
Village head (yes=1)	0.04	0.206	−0.009	0.885
Born in the area (yes = 1)	0.63	0.483	−0.052	0.372

Generally, it can be stated that the shorter the time lag, the higher the innovativeness; however, shorter time lags do not automatically imply higher innovativeness levels, since differences in economic endowments may happen to be the major reason for these differences. To this end, the correlation between time lag and all regression variables was analysed before time lag could be used as an indicator. It was found that economic wealth (liquidity per capita) has no significant correlation with time lag. Farm size as well as the percentage of land under high value crops and under litchi, have a weak but significant correlation with time lag, while variables such as owning a motorbike, and especially exposure and Royal Project membership, correlate much stronger with time lag, and in a highly significant way. These correlations provide evidence of social conformity behaviour among farm households, describing a situation in which adopters usually wait and see until someone else adopts a technology.

A time lag may also be caused by the comparative disadvantage of a new technology when compared to an old technology among farm households. The low level of profitability of an innovation could also explain a prolonged adoption period. A t-test for differences in output per hectare and per month did not reject the null hypothesis of the difference in output between non-GAP certified and GAP certified cropping activities being smaller or equal than 0 (Ho: diff $\leq$ 0). Land-uses for which GAP certification is common are the cultivation of leafy vegetables ($t = 0.807$, $p = 0.201$), greenhouse vegetables ($t = -3.042$, $p = 0.999$) and litchis ($t = 1.045$, $p = 0.176$).

3.3 Model description of the MPMAS Mae Sa watershed application

3.3.1 The methodological context of MPMAS

A multitude of multi-agent systems (MAS) or agent-based models (ABMs) of social, ecological and socio-ecological processes have been developed (An, 2012; Filatova et al., 2013). ABMs in agricultural economics are useful in situations where model complexity leads to analytical intractability, that is, equilibrium conditions either cannot be identified or analytically solved (Nolan et al., 2009).

MPMAS belongs to a category of models referred to as agent-based models of land-use and land-cover change (ABM/LUCC). These models are characterized by the combination of a cellular component representing the physical landscape with an agent-based component representing human decision-making (Parker, 2003). The interactions of autonomous individuals with each other, as well as with the landscape, are important features of ABM/LUCC, which are effective at analysing a variety of resource management problems and which add to the capabilities of standard bio-economic models (Berger et al., 2006).

MPMAS mainly distinguishes itself from alternative models in the ABM/LUCC category, such as Cormas (Becu et al., 2008) for example, through its use of whole farm mathematical programming (MP) to simulate the land-use decision-making of farm households, as the driver of land-use change in agriculture and forestry. Agents can maximize their incomes based on production functions and household expenditure models, and subject to individual resource constraints. The decision-making component is firmly grounded in the micro-economic theory of agricultural economics. While in several ABM applications the rational decision-making of agents is considered as unrealistic, with decision-trees or condition-action rules are advocated instead, Schreinemachers and Berger (2006) argued that the disadvantages of such approaches outweigh the disadvantages of the rationality assumption when modelling complex farm decisions. Nolan et al. (2009) affirmed that agents may be modelled as goal oriented and even as optimizers; however, in a complex system they cannot be considered fully rational. Therefore, the assumption of full economic foresight in MPMAS is relaxed by incorporating adaptive expectation formation and incomplete knowledge through a network model of innovation diffusion. On top of that, the software can be combined with a range of biophysical models to simulate crop yield responses to changes in the crop water supply or changes in soil nutrients.

Altogether, MPMAS allows for the spatially-explicit modelling of human-environment interactions across a wide range of agro-ecosystems and for a variety of purposes (for examples of its applications, see Schreinemachers et al. (2011)). MPMAS can be applied to help understand how the adoption of agricultural technologies or sustainability innovations, how policy intervention and/or how global change processes affect a heterogeneous population of farm households, as well as the economic and environmental resources on which these households rely.

3.3.2 Model set-up and dynamics

The application of MPMAS to assess pesticide use reduction strategies for Thai upland agriculture requires the use of a subset of the full range of input files available in MPMAS. The model is configured to suit the local conditions and the research topic, and comprises the following inputs:

- The agent population, which is subdivided into 15 clusters and which includes information on assets and resource endowments, and agent characteristics.

- Maps, including the spatial representation of agent plots and cluster memberships.
- The MP decision-making component, which is adjusted for each agent during each period.
- The network dynamics component, which defines investment objects and innovation diffusion.
- The perennial crops component, which defines yields and input requirements over the lifespan of a crop, for litchi, roses and IPM vegetables.
- The crop growth component (CropWat model), which specifies crop water requirements.
- Data on the prices of all inputs (pesticides, labour, etc.) and outputs.
- Water rights, a hydrological component regulating water supply (including local weather data), a demographic component defining the labour supply for different age categories, and basic data and a scenario manager, which include general model parameters and switches.

Figure 11 illustrates how the different components of the MPMAS application developed for the Mae Sa watershed are connected. In the model, decision-making depends on market prices, the availability of resources (liquidity, land, assets), the innovativeness of agents and subsequently their access to innovations and, last but not least, a variety of bio-physical information such as plant growth for all crops in the model. The model is recursive, meaning that most factors are updated after each simulation period.

Figure 11: Dynamics of the MPMAS Mae Sa watershed model

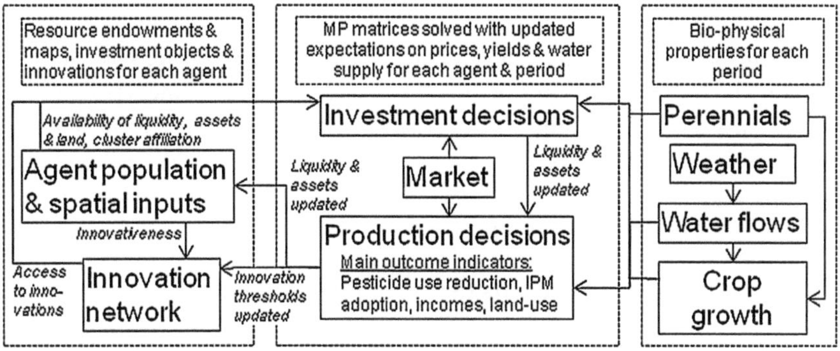

Investment and production decisions in MPMAS are separated into two MP matrices (Berger, 2001), and the acquisition of assets and the adoption of innovations are carried out before cropping choices are made. The investment MP matrix captures the trade-off between short-term income from production and long-term income from investments, as expected net returns averaged over the lifespan of each asset (annuity approach) are optimized simultaneously to production decisions (Schreinemachers et al., 2010). The allocation of cash, labour, pesticides, etc. to a monthly cropping plan occurs after the right-hand side values in the MP matrix are updated for assets and cash, while the annuity values are replaced by actual costs and benefits (Schreinemachers, 2006). The theory of adaptive expectations was incorporated into MPMAS by Berger (2001), so that agents can form expectations on what would happen in the future based on what happened in the past, the development of foresight is in this context not possible. Agents revise their expectations after each simulation period proportional to the difference between actual X_{t-1} and expected vales X^{*}_{t-1}. As long as λ takes values greater than 0 and smaller or equal to 1, agent expectations are adjusted as follows:

$$X^{*}_{t} = X^{*}_{t-1} + \lambda * [X_{t-1} - X^{*}_{t-1}],\ 0 < \lambda \leq 1 \tag{13}$$

The simulation period in the study model is set to five years, in order to reflect the character of the current pest management problems being experienced. The model is based on a cross-sectional dataset, which represents a great variety of cropping activities, meaning it is possible to predict how farmers change their behaviours in response to the incentives and disincentives that affect them in the short-term. However, without some level of knowledge about pest pressure, yields and prices in the medium- and long-term, using an extended simulation period would involve a lot of uncertainty. Furthermore, a perspective on impacts in the short-term is justified, because the issue of pesticide use reduction is on the policy-agenda now and needs to be addressed urgently. Therefore it is relevant to find solutions that can quickly and effectively lower pesticide application levels, plus it is necessary to take into account that farmers place much more importance on the short term risks involved in changing their land-use practices. For these reasons, simulations on the impact of the different pesticide use reduction scenarios are set in the model to run for a period of five years, with three initial spin-up rounds to smoothen the simulation runs.

3.3.3 Asset allocation to create the agent population

In 2010, 1491 farm households were reported as living in the Mae Sa watershed (Schreinemachers, 2010), each of which needed to be represented by a computational agent in the model. Through a survey of 295 farms households, data, as presented in Section 2.2, were collected inter alia on the resource endowments and characteristics of farms in the area, such as labour availability, area under roses or number of greenhouses. A lottery based on Monte-Carlo techniques was used to extrapolate data from the sample of 295 farm households to the population of 1491 farm households. This methodology, the integration of which into MPMAS is described by Berger and Schreinemachers (2006), requires cumulative distribution functions to be used. For a particular resource and agent, a random integer between 0 and 100 is drawn and the quantity of the asset is read from the x-axis. Figure 12 shows examples of cumulative distribution for the greenhouses and areas under litchi area, which, like the area covered by roses, were allocated separately for each slope type.

Figure 12: Examples of cumulative distribution functions

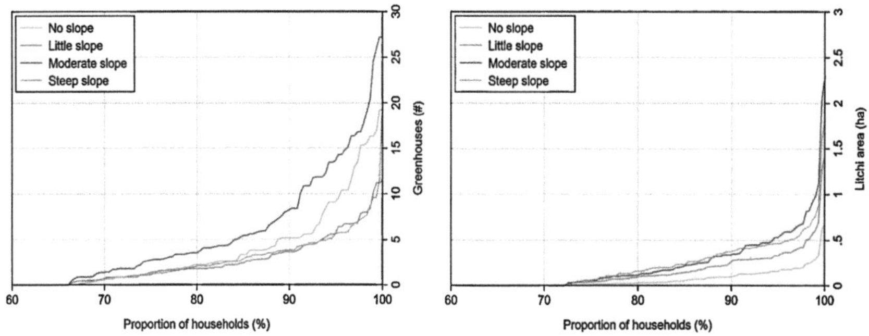

A stochastic element of the Monte-Carlo simulation method is the seed values that are chosen to generate the random population. It is important to create different agent populations by using a variety of seed numbers and then checking for the robustness of results across these populations. Another crucial aspect when initializing the agent population is the relationship between different assets. To take correlations into account, such as between the number of greenhouses and owned land, the sample was subdivided into clusters according to groups of villages and land size. First of all, those villages where cropping habits and suitability due to altitude are similar, were grouped together. The procedure of combining the villages, as listed in Table 9, created five different groups, and each of these groups was then split into three clusters according to the tercile of the farm size (ha).

Overall, this resulted in the creation of 15 clusters within the model, which display a higher degree of homogeneity than the agent population as a whole. The lottery was consequently carried out for each cluster separately, with the outputs of these lotteries merged to produce the agent population. The number of agents in each cluster, their respective IDs (as used by MPMAS) as well as the number of plots owned by agents in a particular cluster, are shown in Table 9 below.

Table 9: Sub-divisions of the agent population

	Group 1			Group 2			Group 3			Group 4			Group 5		
Location	Central watershed			Southern upper watershed			Southern upper watershed			Western upper watershed			Northern upper watershed		
Villages	Pong Yang Nai			Kong Hae			Mae Sa Mai			Buak Tui			Hong Noi Mai		
	Pong Yak Nok						Mae Sa Noi						Hong Noi Gao		
	Muang Kam						Buak Chan								
	Pong Krai														
Ethnicity	Northern Thai			Northern Thai			Hmong			Hmong			Hmong		
Clusters	A	B	C	A	B	C	A	B	C	A	B	C	A	B	C
Cluster IDs	0	1	2	3	4	5	6	7	8	9	10	11	12	13	14
No. of agents	214	216	171	48	36	42	110	118	147	41	39	35	92	88	94
Agent IDs	0–213	214–429	340–600	601–648	649–684	685–726	727–836	837–954	955–1101	1102–1142	1143–1181	1182–1216	1217–1308	1309–1396	1397–1490
No. of plots	332	866	1841	72	156	474	553	1258	2986	164	397	990	575	1002	2160

For the lottery, Stata scripts with a Monte-Carlo mechanism were developed and employed, instead of the internal Monte-Carlo asset allocation mechanism normally used with MPMAS. The agent population created was then read by MPMAS using the so-called direct assignment feature, which requires 'Asset' input files to be developed for each cluster. The direct assignment feature allows for control of the final output of the allocation for each agent, before actual MPMAS simulations are run. In the case of this particular application, it was necessary, having created the whole agent population, to predict the exact innovativeness of each agent using a set of allocated variables representing farm household characteristics and previously estimated regression coefficients (for details see Section 3.3.7).

Figure 13: Process of allocating assets to agents

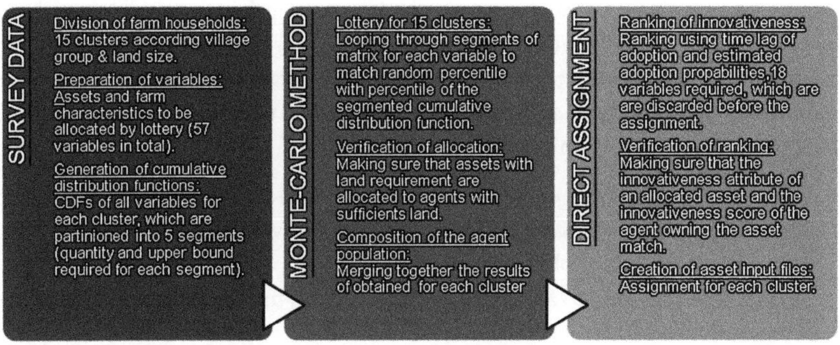

While for each cluster the variables were randomly allocated based on the lottery, innovativeness itself was not assigned in this manner. In fact, the innovativeness ranking procedure carried out for the sample could be transferred as closely as possible to the agent population. Also, minor adjustments could be made for agents possessing greenhouses, an asset not available to adopter group 4, the laggards.

3.3.4 Random spatial allocation of plots and other spatial inputs

The Mae Sa watershed is diverse in terms of land-use, which to a big extent can be explained by its topography. Altitudes ranging from 400 m to 1200 m above sea level, and slopes ranging from very steep to flat expanses, which are mainly located in the central watershed, create a variety of conditions for agriculture. Data collected at the plot level allows the model to distinguish between the topographic properties of these plots. As each empirical cropping observation included in the model was to be associated with a slope gradient, the land available for each agent was subdivided into four slope categories: flat, slight, moderate and steep slopes. Each plot number having its slope type recorded in the survey, a digital elevation model and a map of the available agricultural land were needed.

To randomly allocate the agent plots in the land tenure map, it is possible to use the 'Spatial Randomizer', an executable programme written in C++ and developed at Department 490d at Hohenheim University. This programme allows to generate farmstead and property maps for the whole agent population. Each pixel in these maps corresponds to one plot, the size of which is 0.16 ha.

In this MPMAS application, the random plot allocation is carried out separately for each of the five groups described in Table 9. The five individual land tenure maps are then merged to create the tenure map for the whole watershed. In this way, agents belonging to villages with similar features obtain plots that belong to the agricultural area surrounding these villages. This approach can be considered an example of stratified randomization. Two input files are required to run the Spatial Randomizer, as follows:

- AgentInfo.txt: This file contains information on how many plots each agent owns. The text file was created for each of the five village groups using a Stata script. The number of plots in each of the four slope types, as recorded in the survey, was allocated to agents by the lottery. This information is presented in a separate column for each slope type sorted by the agent ID.
- MapInfo.txt: This file contains the available agricultural area divided into land types according to slope. A digital elevation model (DEM) of the watershed was classified according to the four slope types defined in the survey, after which the resulting slope type map was combined with a map of the agricultural area to produce a map of the agricultural area by slope type. This map was then divided into five parts, in line with the area belonging to the five village groups. The maps were created by using ArcGIS and then exported to ASCII (.txt) format. A map of the agricultural area was then generated by using optical identification of the cultivated land based on Landsat satellite images downloaded in 2010.

The final outputs of the Spatial Randomizer consisted of five land tenure maps (LandTenure.txt), where the pixels representing the plots of a particular agent carried the ID of that agent. These individual maps were put together to obtain the final property map, representing the plots of all agents (CatchMap00Prop.txt), as well as the farmstead map (CatchMap00Farm.txt) on which one plot per agent was assigned to represent farmsteads. Figure 14 shows the property map, with each colour representing the plots of agents belonging to one of the five village groups, while steepness of the slope is indicated by a greyscale elevation model; dark grey being very steep slopes.

Apart from the property and farmstead maps, a population map (CatchMap00Pop.txt) identifying the overall farmland for a population within the map of the study area (only one population was used here) was required as a spatial input for MPMAS. The catchment map (CatchMap00Catchment.txt) similarly defines the overall land available for agriculture and is like all other maps a requirement for running the model. The slope type map of the study area (CatchMap00Soil.txt), produced from a DEM and already employed to create the MapInfo.txt files, was needed to define the suitability of the land for different cropping activities Another spatial input was a map representing the hydrological sectors (CatchMap00Sector.txt) used in the MPMAS routing module. The last two maps required by MPMAS contain information on the cluster membership (15 clusters here) and network membership (one network) for the pixels of the farmsteads (CatchMap00Clu.txt, CatchMap00Netw.txt).

Figure 14: Map of slopes and agent plots in the Mae Sa watershed (each colour represents one group of villages)

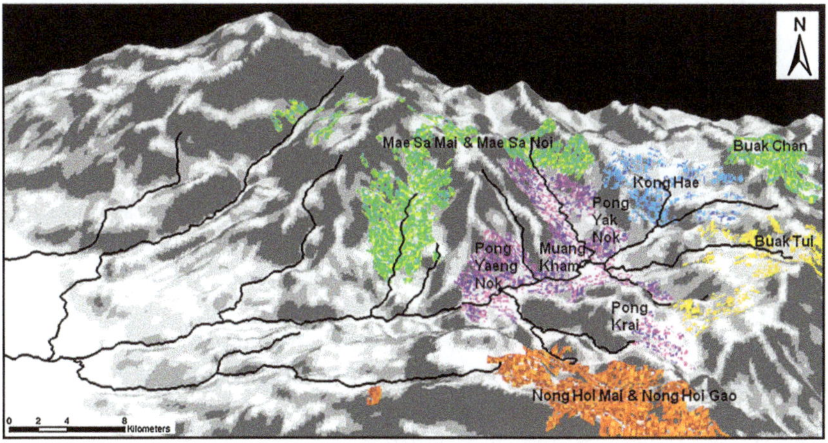

3.3.5 The decision-making component

The decision-making of farm households was modelled according to standard micro-economic theory, using production functions and recursive mathematical programming (MP). As a first step, the cropping activities for the MPMAS model needed to be parameterized. For this purpose, empirical observations from the survey were directly included in the MP matrix. The parameterization approach that will be described in the following section produced 513 different cropping activities, for which a total of 82 active ingredients were needed.

For crops with too few observations to estimate production functions, and which could not be easily grouped together with other crops to obtain a larger number of observations, all data points between ± one standard deviation were selected (for maize, rice, chayote and chrysanthemums). For the perennial crop litchi, cluster analysis was used to generate three management options with different input levels, while the perennial rose, due to there being fewer observations, was inserted with just one average management option. For the remaining vegetable crops, it was possible to group similar vegetables together and to obtain sufficient observation numbers to estimate production functions. Estimations were carried out across three groups with similar pest management, input levels and growing lengths, these being leafy vegetables (open field system), greenhouse vegetables (closed system) and onions/beans (open field system). The production functions helped served as a means to identifying the empirical vegetable data for the model. All observations situated between the upper and lower confidence intervals of predicted output were selected as inputs into the MPMAS model. The Cobb-Douglas production function with a exponential damage control specification for pesticides (see Equation 4), which had already been tested for the quantification of pesticide overuse (Grovermann et al., 2013) - was used to parameterise vegetable cropping activities. This specification allowed the model to take into account the abatement effect of pesticides when defining the agent decision making attributes. The damage control approach and variables used are explained in more detail in Section 3.1.2. It was considered preferable to use a standard Cobb-Douglas production function, because such a function has been found to overestimate pesticide productivity (Lichtenberg and Zilberman, 1986; Praneetvatakul et al., 2003).

For the econometric estimation of production functions for each of the three groups, care was taken to introduce indicator variables alongside growth-stimulating inputs and damage control agents (pesticides), in order to control for farm and individual crop characteristics. For each crop, the output as well as confidence intervals could thus be predicted from the estimated coefficients for the vegetable group, and then be adjusted by the crop coefficient. The values were not only predicted for different levels of pesticide use, but also taking into account the different levels of the other variables used. Observations beyond the upper and lower bounds of the confidence intervals, as displayed in

Figure 15: Representation of the estimated confidence intervals (CIs) used for data selection

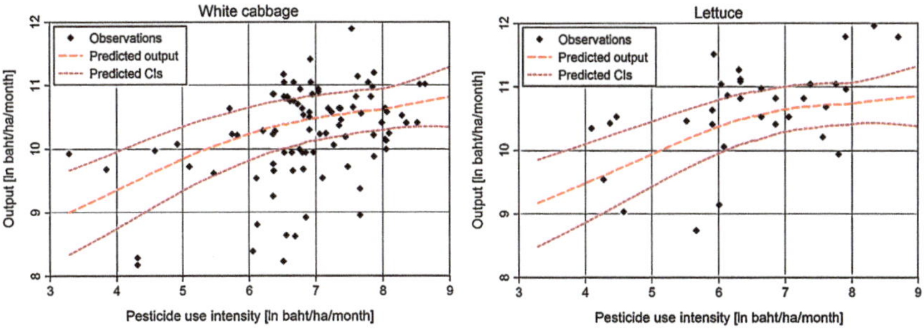

orange in Figure 15 for white cabbage and lettuce, denote the 'outliers', which were excluded. Observations for white cabbage, Chinese cabbage, Chinese kale, lettuce, bell peppers, tomatoes, beans and onions, those between the bounds estimated for the different groups, were included in the MP decision-making matrix.

A constrained MP optimization problem sits at the core of MPMAS, and defines the behaviour of the agents. Agents maximize their farm incomes by selecting an optimal combination of crops based on expectations about prices and yields, and satisfying a large set of resource constraints. Yield expectations, resources, as well as the access to technologies, are updated for every agent at every time step in the model run. Agent decisions are simulated on an annual basis.

Mathematically, the MP can be described as follows:

$$\text{Maximize } f(X) = \sum_{i=1}^{n} p_i x_i \tag{15}$$

$$\text{Subject to } \sum_{i=1}^{n} \sum_{j=11}^{m} b_{ij} x_i \leq y_j \tag{16}$$

$$\sum_{i=1}^{n} x_i \geq 0 \tag{17}$$

Net household income is represented by *f(x)*, which is a linear function of a variety of farm and non-farm activities x_i and their expected return per unit p_i. The farm activities used in the model are summarised in Table 10. The MPMAS solving algorithm selects values for x_i that maximize net household income (Eq. 15) by taking into consideration the various resource requirements of these activities b_{ij} as well as the resource availability y_j (Eq. 16), and by making sure that negative values are impossible (Eq. 17). The coefficients b_{ij} determine the technical requirements, such as pesticide application amounts or labour inputs for example, needed to produce one unit of a certain activity. The sum products of the activities x_i and the coefficients b_{ij} are constrained to be smaller or equal to the right-hand side (RHS), which is denoted by the coefficients y_j. The RHS defines each agent's resource endowments, such as the availability of labour, liquidity and land. Based on these three equations, agents can make their decisions, such as choosing a crop mix with different inputs and outputs, investing in greenhouses or adopting a new technology. Technical coefficients in the decision-matrix, such as on expected yields, available resources or access to particular technologies, are adjusted for each agent and time step. Therefore, the optimisation problem needs to be updated with information stored in other input files, and is repeatedly solved. The Mae Sa watershed MPMAS model requires that the Mixed Integer Linear Programming (MILP) matrix is solved for the investment and production stages of the farm-level decision-making processes of the different 1491 agents for each period in one five year simulation run. In addition, the matrix is solved for the marketing and harvest stages of the decision-making process, in which all activities apart from compensation payments are fixed and thus remain unaltered. The role of compensation payments and the scenarios in which these are used, are explained in Section 3.3.12 and Section 3.4.1. Apart from these particular scenarios, the two extra matrices that are solved do not affect the results, meaning the model set-up actually corresponds to the solving of the MILP matrix over two stages only.

The MP model shown in Table 11 is a reduced representation of the complete decision-making matrix, which contains 1129 columns and 862 rows. For bell peppers and tomatoes, the greenhouse investment and transfer activities are implemented as integers; they are more permanent and stable structures. However, the greenhouses used for growing chrysanthemums are simple, open wooden structures covered by plastic sheeting, and are thus only implemented as per hectare investment and transfer activities. Whereas the permanent greenhouses are allocated by the lottery and entered in the RHS of the matrix for each slope type and then transferred, the makeshift chrysanthemum greenhouses available are entered as a single constraint. Agents can have a maximum of 40 permanent greenhouses, but cannot invest in more than 20 at a time. For perennial crops, constraints are specified for each management option and slope type. The permanent greenhouses and perennial crops need to be fully used, since they are implemented as equal-equal constraints, but idle activities allow agents to keep the greenhouses and perennial crops unmanaged. Similarly, all land needs to be used, but monthly fallow activities allow for not managing the land.

Table 10: Crop data selected for the MP matrix (standard deviations in brackets)

Crop	Observations land 1–4				Growing length	Yield	Labour	Variable inputs	Pesticide use	Irrigated
	(#)	(#)	(#)	(#)	months	(1000 kg/ha)	(mandays/ha)	(1000 baht/ha)	(kg/ha)	(%)
Upland rice	3	2	2	2	5.73	1.68	15.61	2.04	0.20	0
					(0.47)	(0.95)	(7.49)	(1.70)	(0.37)	
Maize	1	2	4	4	4.73	2.24	9.79	0.59	0.07	9
					(0.47)	(0.56)	(6.83)	(1.09)	(0.23)	
White cabbage	6	14	26	20	3.36	25.64	95.19	30.55	1.58	26
					(0.49)	(15.41)	(73.24	(18.30	(1.42)	
Chinese cabbage	6	14	26	20	2.49	23.66	97.84	21.17	2.09	56
					(0.50)	(16.77)	(61.08)	(9.72)	(2.52)	
Chinese kale	8	3	8	4	2.32	5.93	211.83	15.91	1.62	96
					(0.48)	(5.96)	(249.60)	(14.39)	(1.53)	
Lettuce	2	9	10	3	2.48	9.45	100.12	23.68	0.94	54
					(0.51)	(5.80)	(60.66)	(10.19)	(1.48)	
Bell pepper	32	13	6	4	5.64	45.99	247.02	434.08	29.45	100
					(1.16)	(71.21)	(209.21)	(274.93)	(38.01)	
Tomato	2	4	7	2	5.20	68.68	416.18	591.52	14.34	100
					(1.32)	(41.95)	(345.33)	(481.13)	(12.75)	
Onion	3	3	3	3	4.00	26.30	165.29	85.72	8.61	100
					(0.00)	(8.95)	(89.74)	(31.96)	(6.23)	
Green bean	5	11	8	4	3.00	8.97	152.06	13.25	1.59	89
					(0.00)	(8.46)	(102.28)	(12.03)	(1.63)	
Chayote	10	9	2	6	6.23	18.67	178.99	85.76	0.09	96

Crop	Observations land 1–4				Growing length	Yield	Labour	Variable inputs	Pesticide use	Irrigated
	(#)	(#)	(#)	(#)	months	(1000 kg/ha)	(mandays/ha)	(1000 baht/ha)	(kg/ha)	(%)
					(1.37)	(12.83)	(125.40)	(66.38)	(0.38)	
Chrysanthemum	12	2	11	0	4.33	52.12	198.85	32.91	12.23	100
					(0.48)	(35.33)	(244.50)	(15.26)	(8.63)	
Roses	2	2	2	2	12.00	164.37	133.46	131.92	19.44	100
					(0.00)	(147.47)	(69.83)	(59.29)	(11.01)	
Litchi	12	12	12	12	12.00	4.05	11.80	3.32	0.82	67
					(0.00)	(2.84)	(8.81)	(2.89)	(1.04)	

Overall, labour availability for each agent is regulated by the lottery and entered into the RHS of the matrix. Agents can choose to perform off-farm labour and are remunerated with 70 Baht per man day. The extent of off-farm labour is, however, constrained by an off-farm labour constraint, with amounts in the RHS also being allocated to farm agents by the lottery. All agents in the model have access to hired labour, which they can hire on a monthly basis for 96 Baht per man day. Individual buying activities and constraints are specified for all active ingredients, with the objective function in 1000 Baht per kg. Pesticide expenditures are collected in balance rows for the six WHO toxicity classes: WHO 1a, WHO 1b, WHO 2, WHO 3, WHO – unlikely to cause harm and WHO – not listed, and are collected separately for bio-pesticides. For scenarios involving pesticide taxes, the taxes are collected based on the toxicity class. All other inputs (seeds, fertilizers, planting materials and hormones, among others) are aggregated to the balance row variable costs, and are expressed in 1000 Baht per hectare. The balance rows for pesticides, bio-pesticides, pesticide taxes and variable costs are transferred to the capital use balance row, which maintains an equilibrium between the liquid means allocated to each agent and the farm expenditures. Agents can also buy their pesticides and pay the variable costs using credit, to which all the agents have access. The credit limit per agent for each simulation period is 200,000 Baht. Last but not least, the land available under drip and sprinkler irrigation is allocated by the lottery and specified for each cropping activity.

Land, labour, water and sprinkler constraints for the cropping activities are set as monthly, so that different crops can be grown on the same plot in a year, based on an efficient resource use.

3.3.6 Investment objects and innovation diffusion

The decision-making of agents in the model is not only constrained by their resource endowments, but also by their access to investment objects. The parameters, such as acquisition costs or the lifespan of each investment object, as well as the availability of and accessibility to investment objects, are defined in the network module. The investment objects required for the Mae Sa watershed model included simple greenhouses used to grow chrysanthemums, more permanent greenhouses used for growing bell peppers and tomatoes, different perennials (such as litchis, roses and IPM), general short term credit and short term credit for investment in greenhouses, as well as off-farm labour, hired labour, and drip and sprinkler irrigation. Greenhouse objects for bell peppers and tomatoes are distinguished by slope type and whether the investment is made using credit or not. The four slope types, as well as the different input

Table 11: Simplified matrix overview of the MP decision-making model applied to pesticide use reduction strategies in northern Thailand

	Units	# of constraints	Sell crop output	Sell IPM Produce	Off-farm labour	Hire labour	Buy pesticides	Buy biopesticides	Pesticide tax	Buy var. Inputs	Buy pesticides transfer	Buy biopesticides transfer	Buy var. inputs (credit)	Buy pesticides transfer (credit)	Buy biopestic. transfer (credit)	Pesticide taxes transfer	Short-term deposits	Liquid means transfer	Invest veg. greenhouses	Invest veg. greenhouses (credit)	Veg. greenhouse transfer	Veg. greenhouse idle
Units			**kg**	**kg**	**md**	**md**	**kg**	**kg**	**kg**	**baht**	**Baht**	**baht**	**baht**	**baht**	**baht**	**baht**	**baht**	**baht**	**#**	**#**	**#**	**#**
# of activities			**336**	**3**	**1**	**12**	**80**	**5**	**80**	**1**	**6**	**1**	**1**	**6**	**1**	**6**	**1**	**1**	**4**	**4**	**4**	**4**
Objective function			E(+C)	E(+C)	E(+C)	E(–C)	E(–C)	E(–C)	E(–C)	–1	0	0	–1.09	–0.09	–0.09	0	0.02	0				
Investment land	ha	4																	0.02	0.02		
Liquid means	baht	1																1				
Labour	md	1			1																	
Land	ha	4																			0.02	0.02
Monthly water	l/sec	12																				
Perennials land	ha	16																				
IPM land	ha	12																				
Veg. greenhouses	#	4																	(+A)	(+A)	1	1
Land veg. greenhouses	ha	48																			–0.014	
Land chrysanth.	ha	1																				
Land chrysanth. transf.	ha	15																				
Innovation: Credit access	–	1											(+1)	(+1)	(+1)					(+1)		
": Greenh. access.	–	1																	(+1)	(+1)	(+1)	
": IPM access	–	1																				
": perennial access	–	7																				
Land sprinkler irr.	ha	1																				
Land drip irr.	ha	1																				
Off-farm labour	md	1			1																	
Value of var. input	baht	1								–1			–1								(+C)	
Value of pesticides	baht	6					E(+C)				–1			–1								
Value of biopesti-cides	baht	1						E(+C)				–1			–1							

Invest greenhouse chrysanth.	Chrysanth. greenhouse transfer	Invest sprinkler	Invest drip	Transfer sprinkler	Transfer drip	Grow open field crops	Grow flowers	Grow greenhouse crops	Grow fallow	Invest in perennials (litchi, roses)	Grow perennials (Litchi, roses)	Switch growing practices per.	Perennials idle	Invest in IPM vegetables	Grow IPM vegetables	Switch growing practices IPM	IPM plots idle	Labour perennials	Labour IPM	Transfer labour	Transfer land	Transfer annuals	Maximum credit	Maximum greenhouses	Sign	RHS
ha	ha	ha	ha	ha	ha	ha	ha	ha	ha	ha	ha	ha	ha	ha	ha	ha	ha	md	md	md	ha	ha	baht	#		
1	1	1	1	1	1	239	25	70	12	16	16	24	16	12	12	24	12	2	1	1	4	4	1	2		
										1				1											≤	0
																									≤	(R)
																				1					≤	(R)
																					1				=	(R)
						(+A)	(+A)	(+A)		(+A)	(+A)	(+A)		(+A)	(+A)	(+A)									≤	(R)
											1	1													=	(R)
															1	1									=	(R)
																									=	(R)
								1																	≤	0
(+A)	1																								=	(R)
	−1						1																		≤	0
																									≤	(I)
																									≤	(I)
														(+1)	(+1)	(+1)									≤	(I)
										(+1)	(+1)	(+1)													≤	(I)
		(+A)		1																					≤	(R)
			(+A)		1																				≤	(R)
																									≤	(R)
	(+C)					+A	+A	+A																	≤	0
																									≤	0
																									≤	0

	Units	# of constraints	Sell crop output	Sell IPM Produce	Off-farm labour	Hire labour	Buy pesticides	Buy biopesticides	Pesticide tax	Buy var. Inputs	Buy pesticides transfer	Buy biopesticides transfer	Buy var. inputs (credit)	Buy pesticides transfer (credit)	Buy biopestic. transfer (credit)	Pesticide taxes transfer	Short-term deposits	Liquid means transfer	Invest veg. greenhouses	Invest veg. greenhouses (credit)	Veg. greenhouse transfer	Veg. greenhouse idle
Units			**kg**	**kg**	**md**	**md**	**kg**	**kg**	**kg**	**baht**	**Baht**	**baht**	**baht**	**baht**	**baht**	**baht**	**baht**	**baht**	**#**	**#**	**#**	**#**
# of activities			**336**	**3**	**1**	**12**	**80**	**5**	**80**	**1**	**6**	**1**	**1**	**6**	**1**	**6**	**1**	**1**	**4**	**4**	**4**	**4**
Value of pesticide taxes	baht	6							E(+C)							-1						
Capital	baht	1				E(+C)				1	1	1				1	1	-1	(+C)	(+C)		
Maximum credit	baht	1											1	1	1							
Bal. m. sprinkler irr.	ha	12																				
Bal. drip irr.	ha	1																				
Balance pesticides	kg	80					-1															
Balance biopesti-cides	kg	1						-1														
Balance pesticide tax	kg	6							-1													
Bal. monthly labour	md	12				-1													(+A)	(+A)	(+A)	
Bal. labour perennials	md	2																				
Bal. labour IPM veg.	md	1																				
Balance monthly land	ha	48																				
Balance annuals	kg	4																				
Balance rotation	kg	48																				
Balance open field crops	kg	239	1																			
Balance flowers	kg	25	1																			
Balance greenh. crops.	kg	70	1																			
Maximum greenhouses	#	2																	1	1	1	
Balance perennials	kg	4	1																			
Balance IPM veg.	kg	6		1																		

Note: E = Expected values, C = Price coefficients, Y = Crop Yields, A = Technical coefficients, R = Available resources, I = Available innovations. Values in round brackets are adjusted inside the model. Bold values are agent-specific. Values in square brackets mean that irrigation is only used on part of the activities, while others are rainfed.

	Invest greenhouse chrysanth.	Chrysanth. greenhouse transfer	Invest sprinkler	Invest drip	Transfer sprinkler	Transfer drip	Grow open field crops	Grow flowers	Grow greenhouse crops	Grow fallow	Invest in perennials (litchi, roses)	Grow perennials (Litchi, roses)	Switch growing practices per.	Perennials idle	Invest in IPM vegetables	Grow IPM vegetables	Switch growing practices IPM	IPM plots idle	Labour perennials	Labour IPM	Transfer labour	Transfer land	Transfer annuals	Maximum credit	Maximum greenhouses	Sign	RHS
	ha	ha	ha	ha	ha	ha	ha	ha	ha	ha	ha	ha	ha	ha	ha	ha	ha	ha	md	md	md	ha	ha	baht	#		
	1	1	1	1	1	1	239	25	70	12	16	16	24	16	12	12	24	12	2	1	1	4	4	1	2		
																										≤	0
	(+C)										(+A)	(+A)	(+A)		(+A)	(+A)	(+A)									≤	0
																								−1		≤	0
					−1		[1]	[1]	[1]		[1]	[1]	[1]		1	1	1									≤	0
						−1	[1]	[1]	[1]																	≤	0
							+A	+A	+A		+A	+A	+A													≤	0
															+A	+A	+A									≤	0
							+A	+A	+A		+A	+A	+A													≤	0
	(+A)	(+A)					+A	+A	+A										+A	+A	−0.083					≤	0
											(+A)	(+A)	(+A)						−1							≤	0
															(+A)	(+A)	(+A)			−1						≤	0
							1	1	1	1	1	1	1	1	1	1	1					−1				≤	0
							1	1	1														−1			≤	0
										−1													0.42			≤	0
							E(−Y)																			≤	0
								E(−Y)																		≤	0
									E(−Y)																	≤	0
																									−1	=	0
												E(−Y)	E(−Y)													≤	0
																E(−Y)	E(−Y)									≤	0

levels, create a range of investment objects for the perennials (12 for litchi, four for roses and 12 for IPM vegetables).

MPMAS is able to distinguish between investment objects that are innovations and those that are not. A third group of objects is 'symbolic' objects, which are innovations granting access to services and information, but do not usually involve any investment costs (e.g. short term credit). Rogers' original model of technology diffusion, extended by network thresholds and the classification of agents into adopter groups, has been incorporated into MPMAS (Schreinemachers and Berger, 2011), and is used in many applications (Berger et al., 2007; Quang et al., 2014; Schreinemachers et al., 2007b). All objects are listed in the 'Innovation Network', as presented in Table 12, while information per segment is only needed for real innovations, as shown in Table 13. The 'Innovation Network' controls the availability of and access to investment objects. In accordance with the adopter categories and network thresholds defined in Berger (Berger, 2001), the network is split into five innovation segments with the segment IDs 0, 1, 2, 3 and 4. This corresponds to the five adopter categories (innovators, early adopters, early majority, late majority and laggards) used by the theory of innovation diffusion (Rogers, 2003; Valente, 2005).

For each segment, the availability of and access to an object can then be exogenously specified. *Availability* determines when an innovation becomes available to the adopter group that matches the particular segment. For example, innovations may be introduced in year zero of the simulation or later. To exclude an innovation from the simulation run, availability can be set to a prohibitively high value, so that if an investment object is not made available to agents, then access to that object is automatically denied for the given simulation period. *Accessibility* to an investment object is directly related to the exposure level and network threshold attached to that object for a specific simulation period. Only if exposure levels in the agent population reach a segment's network threshold, that is all agents belonging to the previous segments (adopter groups) have adopted, accessibility in that segment switches from '0' (= no) to '1' (= yes).

The network, on top of specifying parameters, such as lifetime or minimum investment size, relates each object to a particular innovativeness level. If an agent possesses an object with an innovativeness level that is greater than his original innovativeness score, the latter is replaced by the former, which might change the initial innovativeness level assigned by the lottery. To this effect, the assignment is interconnected with the allocation of assets by the lottery. In this application, since the innovativeness assignment is under the control of the modeler, it was ensured that an agent with a greenhouse could not be classified as a laggard, because of the innovativeness properties of the greenhouse asset. The required adjustment was anyway only minor. As shown in Table 13, the innovativeness level pertaining to an object should be reflected by the availability and access levels specified for each segment. Ultimately, this specification determines the adoption process stage for an object, and regulates which

Table 12: Example of objects in the network

Object	Type	Innovation ID	Lifetime	…	Innovativeness
Credit	2	−1	100	…	4
Sprinkler	1	−1	10	…	4
…	…	…	…	…	…
Litchi	0	−1	30	…	4
Greenhouse	0	1	15	…	3
IPM veg.	0	2	6	…	0

Note: Type 0 refers to objects with land demand, type 1 to objects without land demand and type 2 to symbolic objects; Innovation Ids are assigned in ascending order, if object is not an innovation, then value is −1.

Table 13: Innovation segments

Segment 0	Availability	Access
Greenhouse	0	1
IPM vegetables	0	0
Segment 1	Availability	Access
Greenhouse	0	1
IPM vegetables	0	0
Segment 2	Availability	Access
Greenhouse	0	1
IPM vegetables	0	0
Segment 3	Availability	Access
Greenhouse	0	1
IPM vegetables	0	0
Segment 4	Availability	Access
Greenhouse	0	0
IPM vegetables	0	0

agents, consistent with their innovativeness score, are next in line to consider adoption.

Apart from greenhouses and IPM, all other objects in the Mae Sa watershed model are not considered as innovations, since farmers commonly know about them and have widely adopted them. As Table 12 shows, litchi and sprinklers for example are not considered innovations anymore, because they are accessible to all agents in the farm population (innovativeness is at the lowest level of '4'). Greenhouses have already been adopted by all agent segments, except the last 'laggard' segment (innovativeness at the second lowest level of '3'), while IPM, in contrast, has not yet been adopted by agents in any segments (innovativeness level corresponds to '0').

The actual technology adoption process in MPMAS consists of two steps (Schreinemachers et al., 2009). As a first step, each agent assesses whether the level of overall exposure to the innovation and the related threshold level match its individual innovativeness, which is determined by which of the five adopter groups agents belong to.. If the threshold is reached, during a second step an agent gains access to the innovation and includes it in the decision-making process – the MP matrix. As illustrated in Figure 16, exposure to an innovation increases as more adoption thresholds are passed and access increases among further adopter groups. This process progressively permits agents to select innovations and if profitable, use them on a specific farm. The profitability of an innovation is evaluated against that of the cropping options existing prior to getting access to the new innovation. Adoption is also subject to constraints, such as the availability of labour, land, cash and other farm assets.

Figure 16: Model of innovation diffusion in MPMAS

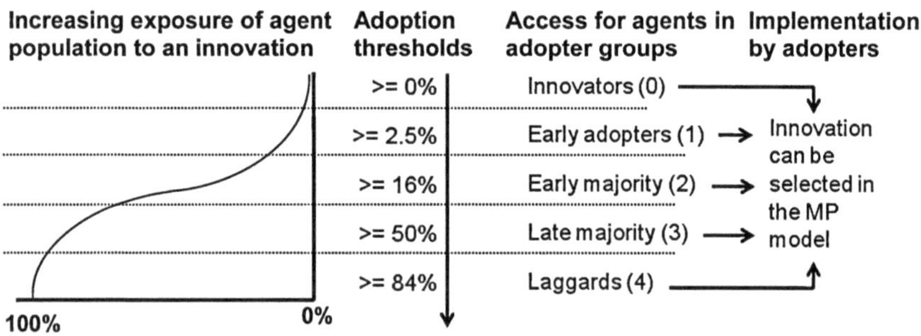

The initial accessibility level establishes how an innovation is further diffused within MPMAS. A completely new innovation such as IPM has an initial adoption figure of 0%; then only agents in the innovator segment (2.5% of the agent population) get immediate access. With an innovation that has existed for a while, such as the greenhouses shown in Table 13, agents in all of the first four segments, and right up to the late majority (84% of the agent population), get immediate access.

3.3.7 Innovativeness ranking and adopter categorisation of agents

As explained above, in the MPMAS model presented here, the adoption time lag and adoption probabilities provide the essential innovativeness characteristics required to assign agents to adopter categories, and the adopter categorisation of agents is based on the estimation procedure and the innovativeness ranking approach outlined in sections 3.2.2 and 3.2.3. The time lag criterion permits an initial innovativeness ranking to be given, which is then further refined by the adoption probabilities. The lottery is used to allocate to all agents the date of adoption (to calculate the time lag) as well as the explanatory variables of adoption listed in Table 8. The coefficients for each of these variables, estimated with the sample selection probit model from equations (8) to (12) and presented in Chapter 5, are used to predict the adoption probability for each agent. The predicted probabilities p_j are given by the formula:

$$p_j = F(x_j * \beta j) \tag{18}$$

Where F is the cumulative normal distribution, x_j is the data vector for the *j-th* observation, and β is the vector of the coefficient estimates.

As described in Section 3.2.3, the agents are ranked, and the cumulative distribution function of this ranking is used to calculate for each agent an exact innovativeness score in accordance with the five adopter groups (0–4). Lastly, the innovativeness scores are cross-checked with the allocation of greenhouses to agents, with those agents possessing a greenhouse but given a laggard score (4) then getting promoted to the late majority adopter group (3).

3.3.8 Perennial crops

Perennial crops are processed differently than annual crops in MPMAS, since the crop yield is not only influenced by input use, but also by crop age. Agents can grow perennial crops on each of the four slope types in the model. The present application includes litchi fruit trees (30 years), IPM leafy vegetable crop rotations (six years) and roses (six years). Three different input-output levels or management options are distinguished for litchi, while for IPM vegetables the distinction is between three different rotation schemes, which is explained in more detail in Section 2.5. These rotations involve a particular cultivation sequence of a number of leafy vegetables and black beans, and vary slightly in terms of input and output amounts. A feature which allows agents the switching of management options, that is, the switching between input-output combinations for litchi, or between rotations for IPM vegetables, is implemented in the Mae Sa watershed model, and guarantees that after an investment in perennial crops has been made, adjustments are still possible for agents during each simulation period. For roses, a flower perennial, only one average input level was included in the model, due to the limited number of observations. A uniform type of management is thus assumed in the model.

Crop yields for perennials are separated between present crop yields and future crop yields. Present crop yields are the expected fruit yields in the current year. If the fruit trees were planted recently, then the present fruit yield is zero and the activity will not be selected when optimizing gross margins in the MP model: the orchard will be left idle until yields become positive. To give agents an incentive to maintain orchards, a second yield row was included in the MP model, this representing the difference between the current yield and the potential or maximum yield. This future yield does add to the optimum solution in the MP model, but does not add to current household incomes nor to agent savings.

In the case of having insufficient funds, agents also have the option to leave their orchard, their IPM plots and their rose fields idle, so receiving no income from these activities. Table 14 shows the most important data required by MPMAS for the perennial cropping activities. Acquisition costs and lifespan figures are defined in the network, while prices are listed under market data and the yields are part of the CropWat model data. The actual perennial input file includes the crop yield factor (a proportion of the crop yield potential provided by the CropWat model), the pre-harvest cash costs, such as fertilising and the labour requirement (labour for management + labour for harvesting). No harvest costs were specified separately within the model,, since all costs are part of the pre-harvest costs and labour is treated separately. Furthermore, the machinery requirement was set to zero, since no machines are used for the perennial cropping activities. In total, the number of perennial crops for which information was included in the input file was 28.

The cash demand in year 0, the expected annual cash demand and the upper bound for short-term credit for perennials, are all calculated internally by MPMAS and inserted for each simulation period into the decision-making matrix using the annualized cash cost, the acquisition cost and some investment parameters.

The leafy vegetables grown with IPM methods are annual crops; however, integrated pest management requires that certain crop rotations are used to break the pest cycle and maintain soil quality. Therefore, several annual cropping activities over the course of one year are grouped together according to what was observed in the field, in order to create three rotation schemes. IPM is knowledge intensive and requires some upfront investments to be made. Light terracing, including the planting of grass strips, as well as simple greenhouse structures which are frequently used to protect the plants against heavy rainfall, make up the acquisition costs.

Table 14: Data of perennials crops in the model

Fruit tree perennial	**Lifespan (Years)**	**Acquisition cost (1000 baht/ ha)**	**Price (1000 baht/ ton)**	**Av. yield potential (ton/ha)**	**Av. crop yield factor**	**Av. labour (mandays/ ha)**	**Av. cash cost (1000 baht/ ha)**
Litchi low input	30	2.43	9.23	2.10	0.78	106.02	0.00
Litchi average input	30	2.43	9.23	3.08	0.78	111.88	2.35
Litchi high input	30	2.43	9.23	5.53	0.78	217.92	6.43
IPM vegetables (specified as perennial crop)	Lifespan (Years)	Acquisition cost (1000 baht/ ha)	Price (1000 baht/ ton)	Av. yield potential (ton/ha)	Av. crop yield factor	Av. labour (mandays/ ha)	Av. cash cost (1000 baht/ ha)
IPM veg. rotation 1	6	9.12	11.70	56.77	0.90	3,378.51	165.4
IPM veg. rotation 2	6	9.12	11.00	64.07	0.90	3,751.65	176.2
IPM veg. rotation 3	6	9.12	12.60	42.86	0.90	3,008.31	130.5
Flower perennial	Lifespan (Years)	Acquisition cost (1000 baht/ ha)	Price (1000 baht/ 1000 flowers)	Av. yield potential (1000 fl. /ha)	Av. crop yield factor	Av. labour (mandays/ ha)	Av. cash cost (1000 baht/ ha)
Roses	6	96.21	8.20	157.64	0.90	1,565.81	141.5

Changing from the conventional production of cash crops with high external input use to integrated practices can be expected to involve some initial yield losses. In the model, this conversion period is taken into account by treating IPM practices as perennial crops and specifying a yield factor for each year. Being a completely new innovation not yet adopted by any agent at the beginning of the simulation, each adopter thus incurs yield losses when starting to grow IPM leafy vegetables. The initial loss of 30% in year one decreases to 20% in year two and then to 10% in year three. From year four onwards, the transition is considered complete and full yields are attained. As mentioned before, IPM in

Table 15: Part of the MP model showing simplified implementation of IPM vegetables as perennial crops

	Units	Constraints (#)	Sell IPM produce	Buy bio-pestic.	IPM vegetables – rotation 1: Invest	Grow	Switch to rotation 2	Idle	IPM vegetables – rotation 2: Invest	Grow	Switch to rotation 1	Idle	Transfer land	Transfer lab.	Sign	RHS
Units			**kg**	**kg**	**ha**	**ha**	**ha**	**ha**	**ha**	**ha**	**ha**	**ha**	**ha**	**md.**		
Activities (#)			**2**	**5**	**12**	**12**	**24**	**12**	**12**	**12**	**24**	**12**	**4**	**1**		
Objective function			E(+C)	E(−C)												
Invest land	ha	4			1				1						≤	0
Land	ha	4											1		=	**(+R)**
Labour	md.	1												1	≤	**(+R)**
Monthly water	l/sec	12			(+A)	(+A)	(+A)		(+A)	(+A)	(+A)				≤	**(+R)**
IPM land	ha	12			+1	+1	+1	+1	+1	+1	+1	+1			=	**(+R)**
IPM innov. Access	–	1			**(+1)**	**(+1)**	**(+1)**	**(+1)**	**(+1)**	**(+1)**	**(+1)**	**(+1)**			≤	**(+I)**
Capital use	Baht	1			(+A)	(+A)	(+A)		(+A)	(+A)	(+A)				≤	**(+R)**
Sprinkler irrigation	ha	12			+1	+1	+1		+1	+1	+1				≤	**(+R)**
Balance bio-pestic.	Kg	5		−1	+A	+A	+A		+A	+A	+A				≤	0
Balance monthly land	ha	48			+1	+1	+1		+1	+1	+1			−1	≤	0
Bal. labour IPM veg.	md.	1			(+A)	(+A)	(+A)		(+A)	(+A)	(+A)			−1	≤	0
Balance IPM veg.	kg	6	+1			**E(−Y)**	**E(−Y)**			**E(−Y)**	**E(−Y)**				≤	0

Note: E = Expected values, C = Price coefficients, Y = Crop Yields, A = Technical coefficients, R = Available resources, I = Available innovations. Values in round brackets are adjusted inside the model. Bold values are agent-specific. md.= mandays the case of upland vegetable production is understood in a strict agro-ecological sense, which creates many similarities with organic agriculture. A variety of examples can be found in the literature on the yield impact of conversion from conventional to organic practices. One comprehensive study comparing the outputs of conventional and organic agriculture found that yield differences are highly contextual, depending on the system used, site characteristics such as soil quality, and the period of time since conversion. The results of this study show a 5% to 34% drop for organic yields (Seufert et al., 2012). Halberg and Silleback (1997) estimated a yield difference of 21 to 37% among grain crops in Denmark. Giovannucci (2006) meanwhile, assessed yield effects during the transition to organic agriculture in China and India. The results of that assessment indicate that production costs are lower, more labour is required and yields initially tend to be 20% to 30% lower on average for organic agriculture, but that by the third or fourth year onward, organic yields often match or even surpass pre-conversion yields. The assumptions used here on yield reduction during the conversion from intensive conventional horticulture to IPM, and the stepwise increasing yield factor then used, are based on the results described above. Table 15 shows how IPM vegetables are contained in the MP model as a perennial crop.

3.3.9 Crop water demand and yields

Crop yields depend on the amount of water supplied, and the CropWat model (Allen, 1998; Doorenbos et al., 1979) and can be incorporated within MPMAS as an optional crop growth tool, as is the case for the Mae Sa watershed application. As with previous applications of MPMAS with respect to land-use decision-making in Thailand (Schreinemachers et al., 2009; Schreinemachers et al., 2010), the present model uses the CropWat model to simulate yield responses to water supply. For each month m and each cropping activity c in the model, the crop water requirement ($CWR_{c,m}$) is specified as a function of potential evapotranspiration (ETO_m), a crop coefficient ($Kc_{c,m}$) and the planted area ($Area_{c,m}$):

$$CWR_{c,m} = ETO_m * Kc_{c,m} * Area_{c,m} \tag{19}$$

The monthly potential evapotranspiration (ETO_m) rate is computed with the CropWat software from FAO (version 8.0), which itself uses the Penman-Monteith formula (Clarke et al., 1998; FAO, 2013a). The required monthly temperatures, humidity, wind and sunshine data were collected from the weather station at the Royal Project site in the watershed. Time-series data was acquired for the period 2005 to 2011, and this was averaged for use in the model. The ETO values per day were multiplied by the number of days in each month and entered into the Crop-

Wat model with the same monthly *ETO* values, as listed in Table 16, and applied to all cropping activities.

Table 16: Meteorological data for the CropWat model

Month	Min Temp °C	Max Temp °C	Humidity %	Wind km/day	Sun hours	ETo mm/day	Rain mm	Eff. rain mm
January	14.9	22.3	73	29	8.5	2.62	7.6	7.5
February	17.5	23.4	58	9	9.1	3.06	1.2	1.2
March	20	24.5	56	23	8.7	3.7	13.8	13.5
April	21.4	25.5	62	14	8.6	4.14	75.4	66.3
May	20.5	24	81	16	7.5	4	184.3	130
June	20.5	23.8	82	5	5.8	3.54	180.6	128.4
July	20.4	23.3	85	8	4.9	3.31	162.7	120.3
August	20.2	23.3	87	9	4.8	3.23	193.4	133.6
September	20	23	88	32	5.6	3.22	246.9	149.4
October	19.2	22.8	87	4	6.9	3.14	140.7	109
November	16.8	21.7	83	9	7.7	2.73	38.9	36.5
December	14.3	21	81	7	8.2	2.41	6.4	6.3

The *Kc* values in the model were derived from the standard values held within the CropWat software's database for the different crops. Each crop has a different growing length and can be planted on different dates. To make adjustments for each of the growing periods (as shown in Table 17), *Kc* values could be obtained for each cropping activity over 10-day intervals for the initial, development, mid- and late-season stages of crop growth. These figures needed to be converted to monthly values for the crop configurations shown in Table 17, in order to specify the monthly CropWat set-up in MPMAS. For many annual crops, such as cabbages and green beans, rainfed as well as irrigated activities exist, while for bell peppers, tomatoes, onions, chrysanthemums and chayote, irrigation is used. Upland rice is the only crop for which irrigation is not used at all. The irrigation method used for the greenhouse crops is drip irrigation, and sprinklers are used for all the open field crops. Perennial activities (roses, litchi and IPM vegetables), which are grown from January to December over consecutive years, require sprinkler irrigation, apart from the low input litchi management option.

The crop water requirement can either be met through irrigation or rainfall,

Table 17: Annual crop data as specified in the CropWat MPMAS input

Crop	Water supply Type	No. of activities	Growing length (months)	Planting month	Av. Kc value
Upland rice	Rainfed	11	5 & 6	Aug, Sep, Oct	0.330
Maize	Rainfed	10	4 & 5	Oct, Nov, Dec	0.278
	Irrigated	1	4	Oct	0.213
White cabbage	Rainfed	23	3 & 4	Aug, Sep, Oct, Nov	0.236
	Irrigated	16	3 & 4	Jan, Feb, Mar, Apr, May, Jul, Sep, Oct, Dec	0.203
Chinese cabbage	Rainfed	30	2 & 3	Feb, Aug, Sep, Oct, Nov, Dec	0.176
	Irrigated	35	2 & 3	Jan, Feb, Mar, Apr, Aug, Sep, Oct, Nov, Dec	0.171
Chinese kale	Rainfed	1	3	Aug	0.212
	Irrigated	24	2 & 3	Feb, Mar, Apr, Jul, Aug, Sep, Oct, Nov	0.147
Lettuce	Rainfed	9	2 & 3	Aug, Oct, Nov, Dec	0.172
	Irrigated	12	2 & 3	Jan, Mar, Apr, May, Jul, Aug, Oct, Dec	0.176
Bell pepper	Irrigated	53	4, 5, 6 & 7	Jan, Feb, Mar, Jul, Aug, Sep, Oct, Nov, Dec	0.199
Tomato	Irrigated	15	4, 5, 6 & 7	Feb, Mar, Apr, Jul, Sep, Oct, Nov	0.365
Chayote	Irrigated	26	5, 6, 7 & 8	Jan, Feb, Mar, Apr, May, Jul, Aug, Nov, Dec	0.396
Green bean	Rainfed	2	3	Aug	0.188
	Irrigated	26	3	Mar, Apr, May, Jun, Jul, Aug, Sep, Oct, Nov	0.180
Onion	Irrigated	12	2, 3 & 4	Jan, Feb, Mar, Apr	0.281
Chrysanthemum	Irrigated	27	4 & 5	Feb, Mar, Jul, Aug, Sep, Oct, Nov, Dec	0.195

which are both converted by the CropWat software (Clarke et al., 1998; FAO, 2013a) to effective rainfall ($ER_{c,m}$) using the USDA S. C. method. Effective rainfall represents the actual water from rainfall available to a crop, which is shown in the last column of Table 16. The irrigation requirement ($IRR_{c,m}$) for each crop in m^3/sec/ha – which is needed in order to determine the monthly irrigation constraints – is estimated from the difference between the crop water requirement and the effective rainfall:

$$IRR_{c,m} = CWR_{c,m} - ER_{c,m} \tag{20}$$

The amount of water then supplied ($CWS_{c,m}$) to a crop is the sum of irrigation water and effective rainfall:

$$CWS_{c,m} = IRR_{c,m} - ER_{c,m} \tag{21}$$

The crop yield reduction factor (Kr_c) is the ratio between the water supply and the water requirement for each crop, and so identifies any potential water scarcity. It is calculated as follows:

$$Kr_c = \left(\frac{1}{m} * \sum (CWS_{c,m} / CWR_{c,m}) \,|\, CWR_{c,m} > 0 \right) \tag{22}$$

The actual crop yield (Y_c) is a function of the crop yield reduction factor and the potential yield ($YPOT_c$) for Kr_c, greater than or equal to 0.5. The actual yield is reduced to zero if Kr_c has a value below the 0.5 threshold:

$$Y_c = \begin{cases} Kr_c * YPOT_c & \text{if } Kr_c \geq 0.5 \\ 0 & \text{if } Kr_c < 0.5 \end{cases} \tag{23}$$

Altogether, the above formulas allow for the determination of yields for each cropping activity included in the MP decision-making matrix. After each simulation period, the actual yields are inserted into the matrix by MPMAS. The expected yield of a crop, on which the actual choice of agents is based, depends on the expected yield realized in the previous year, and not on the actual yield coefficient inserted in the matrix. Using the so-called 'Crop Similarity' feature for smart yield expectations guarantees that the expected yields of any missing crops can be estimated from the expected yields of similar crops. This prevents farmers from developing erroneous expectations if the actual yields of certain crops they used to grow were zero due to a lack of water being available. Three levels of similarity are implemented in this application, with the first level consisting of 73 clusters, the second 20 and the third 11. If at the first level all yields in a cluster are zero, the yields of the less refined clusters on level two are drawn upon.

3.3.10 Irrigation water supply

The irrigation water availability for each hydrological sector in a month is controlled by the routing module in MPMAS. As explained in the previous section, weather data from the Royal Project station plus the CropWat software were used to calculate effective monthly rainfall, which was assumed constant for each year of the simulation. Within the model, daily precipitation data from the weather station is aggregated, averaged and then converted into monthly effective rainfall.

Figure 17: Actual and effective rainfall, calculated from weather data taken from the Royal Project station in Mae Sa Mai

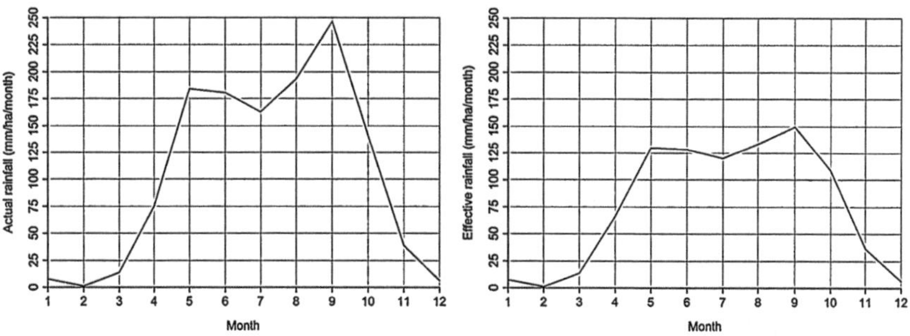

Assuming that there is no water deficit, effective rainfall and groundwater supply are subtracted from the crop water requirement, which results in the creation of irrigation water flows in m^3/second for each sector. The map of hydrological sectors (CatchMap00Sector.txt) defines to which sector the land of a particular agent belongs, and thus the quantity of irrigation water inflow that the agent can expect. MPMAS provides for the inclusion of surface run-off and lateral flows; however, owing to a lack of data, these hydrological aspects were not included in the study application to help calculate the irrigation water inflows for each sector. Instead, inflows are based on effective rainfall and irrigation only. The monthly flows of irrigation water enter the RHS of each agent, representing the irrigation water supply side, while the monthly irrigation requirements, calculated using equation (20), define for each irrigated crop the use of irrigation water in m^3/second necessary to grow one hectare. Rainfed crops do not require irrigation coefficients, since no irrigation constraint applies.

Groundwater pumping is currently only available for some farmers in the central watershed, on ca. 11% of the agricultural land. Cropping with groundwater irrigation (41 activities in the model) was therefore assumed to need no irrigation, as some of the agents in clusters 1–3 have groundwater access. In contrast to groundwater, irrigation water inflows need to be shared in each irrigation sector. The five sectors used for this model correspond to the five groups into which the agent population are subdivided (see Section 3.3.3). Villages with similar topographic features are grouped together, which makes it possible to use the subdivisions as irrigation sectors also. For each household *k* in a sector, the irrigation water supply (IRR_k) is approximated using a backward calculation of irrigation requirements, in line with the approach described by Schreinemachers et al. (2010) in their MPMAS application used for assessing litchi innovations:

$$IRR_k = \Sigma_m \Sigma_c (CWR_{c,k,m} - ER_m - GR_{m,c}) * IEFF_{j,c,k} \quad (24)$$

$IEFF_{j,c,k}$ indicates the efficiency of irrigation method j used on crop c by household k. The results are 0.711 for sprinkler irrigation, 0.8 for drip irrigation and –1 for rainfed crops. This parameter is then used to multiply the effective water supply, which equals the difference between the crops water requirement ($CWR_{c,k,m}$), the effective rainfall (ER_m) and the groundwater ($GR_{m,c}$).

Table 18: Irrigation water supply by month and by sector, in m³/second

Month	Jan.	Feb.	March	April	May	June	July	Aug.	Sep.	Oct.	Nov.	Dec.	Total
Days	31	28	31	30	31	30	31	31	30	31	30	31	365
Flows	m^3	m^3	m^3	m^3	m^3	m^3	m^3	m^3	m^3	m^3	m^3	m^3	m^3
Sector 1	0.055	0.074	0.078	0.062	10.00	10.00	10.00	10.00	10.00	0.045	0.033	0.052	0.041
Sector 2	0.012	0.022	0.027	0.026	10.00	10.00	10.00	10.00	10.00	0.004	0.004	0.011	0.010
Sector 3	0.112	0.184	0.205	0.167	10.00	10.00	10.00	10.00	10.00	0.026	0.018	0.053	0.068
Sector 4	0.079	0.108	0.120	0.056	10.00	10.00	10.00	10.00	10.00	0.057	0.036	0.066	0.047
Sector 5	0.066	0.123	0.136	0.107	10.00	10.00	10.00	10.00	10.00	0.030	0.021	0.055	0.047
Total	0.325	0.511	0.567	0.418	50.000	50.000	50.000	50.000	50.000	0.163	0.111	0.238	252.33

Table 18 show the irrigation water supply for each sector. Since irrigation is only carried out by farmers in the dry and cool seasons (October to April), when the water supply is limited, the availability of supply is only constrained in these months and not in the wet season (May to September). The available irrigation water supply needs to be shared across each sector, which is why the exact amount of irrigation water used is specified for each agent though water rights. In each sector, individual irrigation water use as a proportion of total irrigation water consumption determines the level of water rights available, which sums to one, so that ultimately irrigation water is available to the agent. The water rights for agents were derived from the irrigation water use calculated for farm households in the survey. This information was cloned for the model, in order to obtain the water rights all agents in all sectors.

3.3.11 Farmgate selling, input prices and other input data

Market data for each simulation period is specified in a separate input file. This data is inserted by MPMAS into the objective function vector in the matrix and, together with the solution vector, determines the optimal solution for each agent. It comprises buying prices for inputs such as. hired labour, pesticides and other variable inputs, as well as selling prices for the outputs of each cropping activity. Short-term credits and deposits, just like remunerations for off-farm labour, are also part of the market data. As empirical observations were used to parameterize the model, a distinct selling price is also recorded for most cropping activities. As a result, it was necessary to specify separate selling activities for each growing activity in the model. However, this was only applied to annual crops, because for the perennial litchis and roses – which do not directly correspond to observations but were created with cluster analysis – average selling prices are indicated within the market data. For IPM activities, the average price for each of the three rotations implemented in the model was selected as market data.

Table 19: Summary statistics of farmgate selling prices and input prices used in the model

	Crop	No. of activities	Mean (1000 baht/1000kg)	Standard deviation (1000 baht/1000kg)
Selling prices	Upland rice	9	12.00	0.000
	Maize	11	5.00	0.866
	White cabbage	38	3.50	1.695
	Chinese cabbage	66	3.99	1.642
	Chinese kale	23	11.55	2.911
	Lettuce	24	10.57	5.345
	Bell pepper	55	25.28	11.859
	Tomato	15	17.01	4.785
	Onion	12	10.16	3.006
	Green bean	28	8.91	1.828
	Chayote	27	9.06	1.825
	Chrysanthemum	25	2.46	1.114
	Roses	1	9.23	0.000
	Litchi	1	8.20	0.000
	IPM leafy vegetables	3	11.77	0.802
	Off-farm labour	1	0.07 (1000 baht/ manday)	0.000 (1000 baht/ manday)
Buying prices	Hired labour	12	0.09 (1000 baht/ manday)	0.000 (1000 baht/ manday)
	Synthetic pesticides	80	4.06 (1000 baht/ kg)	4.580 (1000 baht/ kg)
	Bio-pesticides	5	0.250(1000 baht/ kg)	0.177 (1000 baht/ kg)

Table 19 summarizes the main selling and buying prices, plus the interest charged (9%) for buying pesticides and variable inputs with short term credit, and the interest charged for making short term deposits (1.4%), both of which) are included in the vector containing market data. Furthermore, for synthetic pesticides (80 buying activities), the same amount of pesticide tax activities is realized in the model, for which the costs in 1000 Baht per kg are included in the market data. In the price vector, the tax values contain a formula that multiplies the tax rate, provided by the Scenario Manager, by the selling price of each synthetic pesticide. (the tax objective function takes on values of zero in the baseline scenario). Similarly, the five bio-pesticide activities contain a formula that permits the introduction of a subsidy, the actual buying price being multiplied by factors expressed as one plus the percentage of the subsidy. The last policy intervention, which was added to the market data, is a price premium for IPM leafy vegetable produce. Here, the selling prices can be multiplied by a factor that increases the value of the IPM produce.

Due to the rather limited size of the Mae Sa watershed when compared to the overall area under horticulture in northern Thailand, the production output for the watershed is assumed not to affect price formation. Farmgate prices at which agents sell their harvest are thus independent of the harvested amounts for a particular crop. Furthermore, due to the relatively short simulation period of five years, a lack of information on future price trends and the price variations in the model, constant market data was assumed. With the focus on pesticide reduction strategies and no major expected trends related to prices in the next few years (inputs and output prices might slightly increase), it was reasonable to stipulate stable market conditions for the model period.

Another input required to run the Mae Sa watershed MPMAS model consists of a demographic module which regulates aging dynamics. This module is organized according to the age categories of the farm households, which were assigned to each agent by the lottery. Each age category is related to a specific manpower supply, with relatively younger farm household members providing more labour than older members, apart from children up to 14 years of age. From year to year, each agent household member ages by one year, which might change the amount of labour available to the agent.

Finally, a Basic Data file and a Scenario Manager specify general model parameters and switches, which can be used to turn features on or off, and to vary parameters for different simulation runs. Access to innovations, policy intervention levels, the formation of expectations or the random seed value for initializing the agent population, can be varied in this way among scenarios for example. For this research study, the model is used to compare 42 pesticide use reduction scenarios, the set-up of which is explained in more detail below, while the outcomes of the simulation experiments are described in the Chapter 5.

3.3.12 Tax collection and compensation payments

Owing to the increasing tax burden with increasing pollution levels, polluters usually prefer command and control schemes to taxes. In such a context, the maximum level of pollution is set as a regulation; however, regulations create no incentive to reduce pollution further after a target in the form of a maximum pollution limit has been reached. The acceptance of a tax policy among polluters is said to change though, if tax revenues are redistributed to polluters in the form of a lump sum (Buchanan and Tullock, 1975). To maintain the tax disincentive, the tax itself and the reimbursement of tax revenues need to be decoupled. In this study's model, the intended effect of the tax to change the behaviour of polluters remains, while after production has taken place tax revenues can be reimbursed according to a pre-specified criterion. For environmental taxes, redistribution can happen through a uniform lump sum, a differentiated lump sum payment, e.g. proportional to pollution, or through a subsidy for example, for labour through reduced labour taxes (Felder and Schleiniger, 2002). A tax can also be used to fund more sustainable land-use alternatives. The different incentives for IPM, as explained in the section above, represent a means of using tax revenues to subsidise environment-friendly production practices. However, there is no exact match between revenues and expenditures for the area payment, the price premium or the bio-pesticide subsidy. This would also be impractical, since any such policy needs to offer farmers a concrete incentive in the year of production they are able to incorporate any changes into their cropping plan.

However, this is not the case with a standard compensation payment, one used to redistribute tax money. As a result in the Mae Sa watershed MPMAS application, a lump sum differentiated by farm size is used to examine the effect of compensating agents for their tax expenditures. For this purpose, the payment is combined with medium and high proportional taxes, with agents having and not having access to IPM.

In terms of software implementation, the redistribution of tax revenues is handled by an extra agent, which represents the responsible tax authority. Implementation is based on an MPMAS feature which was developed to model the interaction of farm agents with producer organisation agents. For this study, the producer organisation agent was modified to represent a tax collection agent, and in the model this agent aggregates the taxes paid by all farm agents for the pesticides they apply. A look-up table contains references to the solution vector of the tax accounting activities, as well as the compensation payment activity in the MP matrix for each agent. The same table also provides information on the division required to obtain the updated price for each of these activities. As shown in Table 20, the tax revenues, denoted as (T) and separated according to WHO toxicity classes, are entered into the RHS of the tax collector agent's MP matrix and further aggregated to a total tax revenue value. The RHS value (C) is the aggregate solution vector value over all agents for the compensation payment activity, which is zero.

Table 20: MP Matrix of the tax collection agent

	Units	# of constraints	Transfer tax revenues	Total tax revenues (Numerator)	Total agricultural area (Denominator)	Numerator for tax activities	Denominator for tax activities	Sign	RHS
Units			**kg**	**baht**	**baht**	**Ha**	**Ha**		
# of activities			**6**	**1**	**1**	**1**	**1**		
Objective function			0	0	0	0	0		
Tax revenues	baht	6	1					=	(T)
Compensation	baht	1						≤	(C)
Total agr. Area	ha	1			1			=	2212
Numerator	#	1				1		=	0
Denominator	#	1					1	=	1
Total tax revenue	baht	1	(−1)	1				=	0

Note: T = Collected tax revenue, C = Price of compensation. Values in round brackets are adjusted inside the model

The optimal solution for the total tax revenue is the numerator, while the optimal solution for the total agricultural land in the watershed represents the denominator for calculating the tax revenue per hectare. The RHS contains the overall area available for cultivation, which is 2,212 ha. The per hectare revenue corresponds to the price of the compensation payment, which is returned to the MP matrix of each agent. Since this price is supposed to be inserted in a separate activity and not the tax revenue accounting activities, a work-around was required here; the compensation constraint fulfils the purpose of communicating to MPMAS in which activity the price coefficients needs to be inserted, since it is referenced to the total tax revenue and total agricultural land activities. The updated price for the six tax activities needs to be zero, and is simply computed in the tax collector agent's MP matrix by dividing zero by one.

The implementation of the compensation payments requires that two additional MP matrices are solved for each agent for the so-called marketing and harvesting stages. This is necessary in order for the agents to obtain the payment, which is accounted for in their final income statement. Production and other farm decisions, such as hiring labour, are not affected by this and are fixed in the two final stages. Therefore, in the direct tax compensation scenarios the redistribution of the payment is the only change that occurs after production; in all other scenarios, no changes are made after production, meaning the model set-up actually corresponds to a two-stage decision-making process (investment and production).

3.3.13 SWAT-based pesticide use constraints – chlorothalonil and cypermethrin[4]

Whereas the focus of this study is on the assessment of practical policy interventions, those workable under the conditions prevalent in northern Thailand, it will also be interesting to consider interventions that are based more on scientific findings. To this end, simulations were carried out on %-reduction scenarios for two pesticides: cypermethrin – a pyrethroid insecticide, and chlorothalonil – a chloronitrile fungicide. The pesticide fate and transport of both active ingredients is modelled with the Soil and Water Assessment Tool (SWAT), which was calibrated for in the Mae Sa watershed by Bannwarth et al. (2014). Therefore it is possible to evaluate concentrations of chlorothanonil and cypermethrin against pre-defined eco-toxicological thresholds (No Observed Effect Concentration (NOEC) and Predicted No Effect Concentration (PNEC)), as calculated by Sangchan et al. (2013).

A prerequisite for this is that the initial application amounts for the baseline be similar in both models, which is the case. In total, 1460 kg of chlorothalonil and 1100 kg of cypermethrin were applied on a total area of 2095 hectares. The average application rates used, based on the household survey, were 0.80 kg/ha for chlorothalonil and 0.53 kg/ha for cypermethrin, which accounted for 12.5% and 6.6% of the total amount of pesticides applied, respectively. These application rates roughly corresponded to the estimated application rates taken from the SWAT model, which made these chemicals suitable for the study. The eco-toxicological value of cypermethrin is higher than that of chlorothalonil in the model,

4 Parts of this section have been submitted as: Bannwarth, M., Grovermann, C., Schreinemachers, P., Ingwersen, J., Lamers, M., Berger, T., Streck, T., subm. Non-hazardous pesticide concentrations in surface waters: Simulating application thresholds and resulting farm income effects. Journal of Environmental Management.

Table 21: Key physico-chemical properties and application data of chlorothalonil and cypermethrin

Pesticide (common name)	Chlorothalonil	Cypermethrin
Usage type[a]	Fungicide	Insecticide
Structure formula[a]	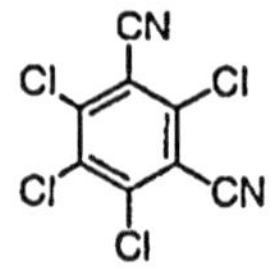	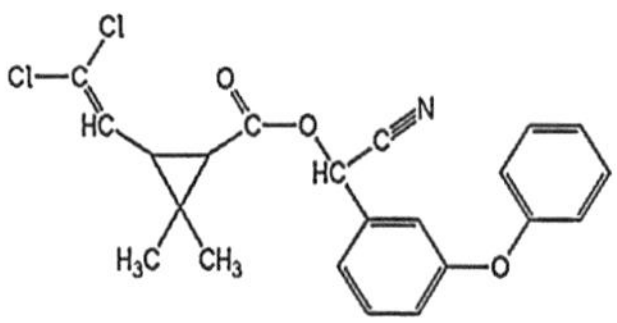
Substance group[a]	Chloronitrile	Pyrethroid
Solubility in water (20°C)[a]	0.8 mg/l	0.009 mg/l
Soil degradation (aerobic, DT50, lab at 20°C)[a]	15.7 days	68 days
Recommended application rate[a]	1–2.5 kg/ha	0.025–0.150 kg/ha
Observed application rate[c]	0.8 kg/ha	0.53 kg/ha
WHO toxicity classification[a]	U (unlikely to cause harm)	II (moderately hazardous)
LC_{50} (median lethal concentration, fish)[b]	38 µg/l	2.8 µg/l
Critical NOEC (no observed effect concentration, fish)[b]	3 µg/l	0.04 µg/l
PNEC (predicted no effect concentration)[b]	0.3 µg/l	0.004 µg/l
Share of total pesticide use on leafy vegetables[c]	1 %	36.5 %
Share of total pesticide use on greenhouse veg.[c]	16.3 %	38.2 %
Share of total pesticide use on flowers[c]	69.0 %	8.4 %
Share of total pesticide use on other crops[c]	13.6 %	16.9%

Notes: [a]Taken from Tomlin (2009) and The Footprint Pesticides Properties Database (2013). [b]Taken from Sangchan et al. (2013). [c]Estimated from own survey data.

and the chemical is classified as moderately hazardous by WHO (Table 21). For cypermethrin, pest resistance and resurgence have been recurring problems in Thailand (Areekul, 2007). In an analysis of pesticide residues in vegetables, cypermethrin has been found to be the most prevalent chemical substance residue (Athisook et al., 2007).

Surface water pesticide loads for these two chemicals were calculated by multiplying measured average daily pesticide concentrations by the corresponding mean daily discharge. The analysis of these water samples was previously described in Sangchan et al. (2013; 2012), for which pesticides were extracted from water samples using solid phase extraction (SPE) and kept frozen until further analysis in a laboratory in Germany. After elution from the cartridges, pesticides were then analyzed using gas chromatography (HP Agilent Technologies, 7890 GC) equipped with an electron capture. Samples with outstanding and high peak concentrations were confirmed by gas chromatography-mass spectrometer, with samples of chlorothalonil and cypermethrin coming out at 58% and 69% respectively. The detection limit was set as 1 ng/l for chlorothalonil and 2 ng/l for cypermethrin.

The impact assessment of reducing the application amounts of chlorothalonil and cypermethrin is based on feeding recommended pesticide applications resulting from the SWAT simulations into the MPMAS model. The eco-toxicological threshold values serve as a basis to determine the recommended application levels that define the scenarios. A comprehensive overview of the different components of the integrated modelling approach is shown in Figure 18.

There was no need to introduce a continuous feedback loop between SWAT with MPMAS. This means that farm decisions about pesticides affect the fate of pesticides in relation to surface water, but there is no reverse causality as pesticide concentrations do not affect farm decision-making. The approach used in the model is to first use SWAT to estimate the necessary reductions in pesticide application rates, then evaluate the impact of these using MPMAS. Simulations with SWAT were repeated until the point had been reached at which concentrations amounts of the two pesticides stayed below the designated thresholds. The reductions on the concentration amounts were translated into %-reductions of application quantities and then implemented as scenarios in MPMAS.

The %-reduction in the use of each of the two pesticides is implemented through a per hectare maximum application constraint. The amount of applied substance from the baseline scenario is divided by the amount of land, which

Figure 18: Overview of the input data and the integrated modelling approach with SWAT and MPMAS

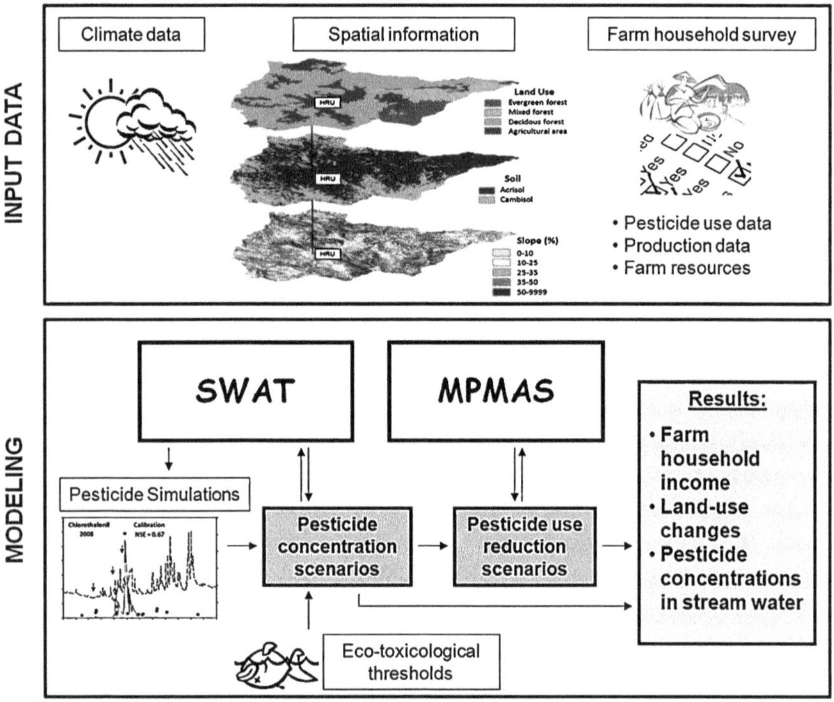

results in an application rate of 0.70 kg/ha for chlorothalonil and 0.53 kg/ha for cypermethrin. In the MP matrix, the land transfer columns contain the application reduction coefficients –*P* for the balance rows 'Max. chlorothalonil' and 'Max. cypermethrin', which besides –*P* contain the technical pesticide application coefficients +*A* for those growing activities requiring the use of chlorothalonil and/or cypermethrin. As can be seen from Table 2, 34 cropping activities require chlorothalonil and other pesticides (grow crops 1), 144 cropping activities require cypermethrin and other pesticides (grow crops 2), seven cropping activities require chlorothalonil, cypermethrin and other pesticides (grow crops 3), while 328 cropping activities require other pesticides only (grow crops 4). The formula for the application reduction coefficient is:

$$-P = -\text{Application rate(kg/ha)} * \text{Reduction(\%)} * \text{Switch}(0|1) \qquad (25)$$

This makes sure that the overall use of chlorothalonil is constrained in accordance with the value of the related application reduction scenarios (50%, 25%, 20%, 18% (PNEC), 15%, 10%, 5%, 2.5%, 1%), and that the overall use of cypermethrin is constrained in accordance with the value of the related application reduction scenarios (50%, 25%, 20%, 15%, 10%, 5%, 2.5% (NOEC), 1%, 18% (PNEC)). As well as the eco-toxicological values (NOEC and PNEC), a range of other reduction scenarios were simulated for the purpose of comparison. We found that the NOEC of chlorothalonil, which is 3μg/l, was never exceeded in the SWAT model. Hence no reduction in the application rate was necessary and it was not used as a scenario.

3.4 Scenario specifications of simulation experiments

In empirical models, any prediction of future changes is usually based on an extrapolation from observed past behaviour and a description of past agricultural technologies. Therefore, such models do not deal easily with specific alternative technological options or new constraints and policies (Falconer, 2001). In contrast, an agent-based mathematical programming model provides the necessary features for a far-reaching ex-ante assessment regarding the impacts of a variety of policy interventions and innovations on a heterogeneous population. MP-MAS simulates the likely responses of farm households to such situations and the changes in resource use that will take place, providing important information for any evaluation and targeting of technology, as well as for policy interventions (Berger et al., 2006).

As mentioned above, the Scenario Manager in MPMAS enables the user to specify changes in parameters required for running a particular scenario. As a result, it is possible to run a sequence of simulation experiments that stepwise isolate the effects of parameter changes (Berger et al., 2006). This means it is possible to design pesticide use reduction scenarios which constrain or affect the decision-making of agents in different ways. For instance, prices for pesticide inputs can be increased, access to IPM can be granted or pesticide use constraints can be switched on or off for different simulation runs. For pesticide use reduction policies, a range of parameters can be used to evaluate the effect of an intervention. This is instrumental to test the sensitivity of results to different levels of change. While not all parameters are reasonable from a policy-making point of view, testing different values allows one to obtain an estimation of the magnitude of change in terms of policy impact in response to increasing or decreasing intervention levels.

Table 22: Implementation of the reduction of chlorothalonil and cypermethrin use in the MP matrix

	Units	# of constraints	Sell crop output	Buy chlorothalonil	Buy cypermethrin	Buy other pesticides	Buy pesticides transfer	Buy pesticides transfer (credit)	Buy other var. inputs	Buy other var. inputs (credit)	Grow crops 1	Grow crops 2	Grow crops 3	Grow crops 4	Liquid means transfer	Land transfer	Maximum credit	Sign	RHS
Units			**kg**	**kg**	**kg**	**kg**	**baht**	**baht**	**baht**	**baht**	**ha**	**ha**	**ha**	**ha**	**baht**	**ha**	**baht**		
# of activities			**341**	**1**	**1**	**78**	**6**	**6**	**1**	**1**	**34**	**144**	**7**	**328**	**1**	**4**	**1**		
Objective function			E(+C)	E(−C)	E(−C)	E(−C)	0	E(−I)	E(−1)	E(−I−1)	0	0	0	0	0	0	0		
Liquid means endowment	baht	1													1			≤	**(R)**
Land endowment	ha	4														1		=	**(R)**
Value of var. input balance	baht	1							−1	−1	+A	+A	+A	+A				≤	0
Value of pesticides balance	baht	6		+C	+C	+C	−1	−1										≤	0
Capital balance	baht	1					1		1						−1			≤	0
Maximum credit balance	baht	1						1		1							−1	≤	0
Chlorothalonil balance	kg	1		−1							+A		+A					≤	0

	Units	# of constraints	Sell crop output	Buy chlorothalonil	Buy cypermethrin	Buy other pesticides	Buy pesticides transfer	Buy pesticides transfer (credit)	Buy other var. inputs	Buy other var. inputs (credit)	Grow crops 1	Grow crops 2	Grow crops 3	Grow crops 4	Liquid means transfer	Land transfer	Maximum credit	Sign	RHS
Units			**kg**	**kg**	**kg**	**kg**	**baht**	**baht**	**baht**	**baht**	**ha**	**ha**	**ha**	**ha**	**baht**	**ha**	**baht**		
# of activities			**341**	**1**	**1**	**78**	**6**	**6**	**1**	**1**	**34**	**144**	**7**	**328**	**1**	**4**	**1**		
Cypermethrin balance	kg	1			−1							+A	+A					≤	0
Max. chlorothalonil balance	kg	1									+A		+A			(−P)		≤	0
Max. cypermethrin balance	kg	1										+A	+A			(−P)		≤	0
Other pesticides balance	kg	78				−1					+A	+A	+A	+A				≤	0
Monthly land balance	ha	48									1	1	1	1		−1		≤	0
Yield balance	kg	341	1								**E(−Y)**	**E(−Y)**	**E(−Y)**	**E(−Y)**				≤	0

Note: E = Expected values, C = Price coefficients, I=Interest rate, Y = Crop Yields, A = Technical coefficients, R = Available resources, P = Pesticide reduction coefficients. Values in round brackets are adjusted inside the model. Bold values are agent-specific.

Table 23: Overview of policies at different intervention levels

Intervention	Low (1)	Medium (2)	High (3)
Flat tax	10%	30%	50%
Prop. tax	WHOIa & Ib: 20% WHOII: 15% WHOIII: 10% WHOU: 5% WHONL: 5%	WHOIa & Ib: 50% WHOII: 40% WHOIII: 30% WHOU: 20% WHONL: 20%	WHOIa & Ib: 70% WHOII: 50% WHOIII: 40% WHOU: 30% WHONL: 30%
Area payment for area under IPM	Year 1: 50,000 baht/ha Year 2: 25,000 baht/ha Year 3: 12,000 baht/ha	Year 1: 100,000 baht/ha Year 2: 50,000 baht/ha Year 3: 25,000 baht/ha	Year 1: 200,000 baht/ha Year 2: 100,000 baht/ha Year 3: 50,000 baht/ha
Price premium for IPM produce	2% price increase	5% price increase	10% price increase
Bio-pesticide subsidy (IPM input)	20% price decrease	40% price decrease	60% price decrease

A wide range of policy interventions can be resorted to achieve reductions in the application of chemical crop protection substances. As outlined in Section 1.4, the focus of this study is on financial instruments and sustainability innovations. Table 23 gives an overview of the different computer experiments carried out to simulate the policy interventions employed for assessing pesticide use reduction strategies within MPMAS, all of which can be simulated stand-alone or as combined interventions. The interventions can occur at different levels, denoted as low, medium and high. More precisely, the Mae Sa watershed model includes options to introduce taxes (flat and proportional) and lump sum compensation payments, as well as subsidies, area payments and price premiums in conjunction with the introduction of IPM technology. Taxes and, in particular, the economic incentive instruments, are evaluated in terms of their capability to reduce pesticides use, their impact on farm incomes, their cost-effectiveness and how far they induce agents to adopt sustainable pest management practices. Access to IPM, which is treated as a complete innovation, is granted to the innovator segment, as specified in the Scenario Manager.

The following sections outline the various policy-oriented pesticide use reduction scenarios, while Section 3.3.13 puts forward a range of scientifically specified scenarios that were designed based on information obtained from the pesticide fate and transport model SWAT. The SWAT-based scenarios are simply named after the pesticide (CHL = chlorothalonil, CYP = cypermethrin) and the percentage of baseline reduction applied. Meanwhile the naming of the pesticide policy scenarios follows a different system to clearly distinguish them. Each policy intervention is abbreviated to three characters. In total, a scenario can consist of 15 characters. The first three characters indicate the baseline scenario or alternatively the different allocations of innovativeness scores to agents, most importantly, the IPM access set-up developed for this application. The subsequent three characters denote the different tax scenarios, while the remaining nine characters relate to the different IPM adoption incentives. Letters refer to the policy instrument, and numbers to the level of intervention (1 = low, 2 = medium, 3 = high). In the tables below, all scenario names are listed and the interventions are explained in more detail.

3.4.1 Pesticide taxes

Scenarios 1–6 (see Table 24) are set up to simulate a series of tax interventions without giving agents access to IPM. While the flat tax applies a uniform disincentive to buy pesticides, the proportional tax is graded according to the hazardousness of the substance. This hazardousness is represented by the WHO toxicity classes, which are further explained in Section 2.4. The proportional tax allows the model to penalize the use of toxic substances more than the use of lower-risk active ingredients. This kind of tax intervention is more useful if the aim is to target the use of undesirable, harmful pesticide products. For each proportional tax scenario, a corresponding tax scenario with a lump sum compensation payment (scenarios 7–9) is simulated, which gives an idea as to how far a redistribution of the collected taxes to the agents can offset their income losses. The tax rates presented in this study follow the logic explained in Section 1.4; they are set to levels that are reasonable from a policy implementation perspective. Two very high flat tax rates (75% and 100%) were also tested for comparison purposes, and these are presented in the Annex.

Table 24: Pesticide tax and tax compensation scenarios simulated by MPMAS

Scenario ID	Interventions	Description	Scenario names
1–3	Flat tax	Flat taxes, stand-alone, 3 tax levels	___TF1_________ ___TF2_________ ___TF3_________
4–6	Prop. Tax	Proportional taxes, stand-alone, 3 tax levels	___TG1_________ ___TG2_________ ___TG3_________
7–9	Prop. tax + compensation	Proportional taxes and lump sum compensation payments, 3 tax levels	___TC1_________ ___TC2_________ ___TC3_________

3.4.2 IPM access and pesticide taxes

As shown in Table 25, scenarios 10 to 12 are set up to test different innovativeness assignment configurations, the resulting diffusion in the innovation network and the adoption of IPM. This is mainly needed for validation/sensitivity analysis purposes. Differences in these three configurations can be summarised as follows:

1) IPM: IPM access to innovators only combined with direct assignment of innovativeness (explained in sections 3.3.3 and 3.3.7).
2) RAN: IPM access to innovators only combined with random assignment of innovativeness using Monte-Carlo techniques.
3) ALL: Direct assignment of innovativeness, but IPM access to all 5 innovation segments [0–4]

Table 25: IPM access and IPM access in conjunction with pesticides tax scenarios simulated by MPMAS

Scenario ID	Interventions	Description	Scenario names
10–12	IPM access	IPM innovation introduced, stand alone, testing different innovativeness configurations	IPM____________ RAN____________ ALL____________
13–15	IPM access + flat tax	IPM innovation introduced & flat tax, 3 tax levels	IPMTF1_________ IPMTF2_________ IPMTF3_________
16–18	IPM access + prop. Tax	IPM Innovation introduced & prop. tax, 3 tax levels	IPMTG1_________ IPMTG2_________ IPMTG3_________
19–21	IPM access + prop. tax + compensation	IPM Innovation introduced & prop. tax & lump sum compensation payments, 3 tax levels	IPMTC1_________ IPMTC2_________ IPMTC3_________

For all policy scenarios, as well as the baseline scenario, the first innovativeness configuration (IPM) is used. At the beginning of the simulation, access to IPM is granted to segment 0 in scenarios 13 to 44. Scenarios 13 to 21 combine the different taxation schemes, excluding and including compensation payments, with IPM access.

3.4.3 IPM access and adoption incentives

While the above scenarios simulate the impact of IPM in conjunction with financial disincentives, the following scenarios combine IPM with one of the financial incentive mechanisms shown in Table 23.

Table 26: IPM access in conjunction with supportive policy scenarios simulated by MPMAS

Scenario ID	Interventions	Description	Scenario names
22–24	IPM access + area payment	IPM innovation introduced & payment per area under IPM, 3 levels	IPM___PA1______ IPM___PA2______ IPM___PA3______
25–27	IPM access + price premium	IPM innovation introduced & price premium for IPM produce, 3 levels	IPM______PR1___ IPM______PR2___ IPM______PR3___
28–30	IPM access + bio-pesticide subsidy	IPM innovation introduced & subsidy for bio-pesticides, 3 levels	IPM_________SU1 IPM_________SU2 IPM_________SU3

In the context of a sustainability innovation, such as IPM, it is important to assess a variety of supportive policies in terms of their impact on the diffusion of the innovation, on pesticide use, livelihoods and land-use. These policies offer alternatives to using the tax revenues in a more targeted fashion, rather than simply redistributing them according to farm size, as simulated by the lump sum payments. The graduated area payment depends on the year of introduction of IPM for leafy vegetables. As outlined in Section 3.3.8, agents adopting IPM practices face a reduced yield during the conversion period from conventional to integrated cultivation practices, with this yield loss gradually declining from 30% in year one to 0% in year four. Therefore, the area payment is high in the first year, but then progressively drops until it reaches zero in year four after adoption. The subsidy for bio-pesticides involves a 20% price decrease for the low intervention, a 40% decrease for the medium intervention and a 60% decrease for the high intervention. The price premium entails price increases of 2%, 5% and 10% respectively.

3.4.4 Policy mixes

The remaining policy scenarios, 31 to 44, consist of a mixture of a tax disincentive and different combinations of the IPM incentive instruments, resorting to the interventions described above. Several additional scenarios were tested, but for a better comprehension of the results only a selection of the more relevant simulation experiments are included in this study. An important criterion for deciding on the relevance of a scenario was, besides effectiveness, the financing. The scenarios presented in Table 27 represent a combination of policies that involve a range of policy costs. For the presentation of the results, a distinction between cheaper and more expensive policy mixes was made.

Table 27: Mixed policy scenarios simulated by MPMAS

Scenario IDs	Interventions	Scenario names
31	IPM access + high prop. tax + low area payment + low price premium	IPMTP3PA1PR1___
32	IPM access + high prop. tax + low area payment + med. price premium	IPMTP3PA1PR2___
33	IPM access + high prop. tax + low area payment + med. bio-pesticide subsidy	IPMTP3PA1___SU2
34	IPM access + high prop. tax + low area payment + high bio-pesticide subsidy	IPMTP3PA1___SU3
35	IPM access + high prop. tax + low area payment + low price premium + low bio-pesticide subsidy	IPMTP3PA1PR1SU1
36	IPM access + high prop. tax + low area payment + low price premium + med. bio-pesticide subsidy	IPMTP3PA1PR1SU2
37	IPM access + high prop. tax + low area payment + med. price premium + low bio-pesticide subsidy	IPMTP3PA1PR2SU1
38	IPM access + high prop. tax + low area payment + med. price premium + med. bio-pesticide subsidy	IPMTP3PA1PR2SU2
39	IPM access + high prop. tax + med. area payment	IPMTP3PA2______
40	IPM access + high prop. tax + med. price premium	IPMTP3___PR2___
41	IPM access + high prop. tax + med. price premium + low bio-pesticide subsidy	IPMTP3___PR2SU1
42	IPM access + high prop. tax + med. price premium + med. bio-pesticide subsidy	IPMTP3___PR2SU2
43	IPM access + high prop. tax + med. price premium + high bio-pesticide subsidy	IPMTP3___PR2SU3
44	IPM access + high prop. tax + high bio-pesticide subsidy	IPMTP3______SU3

Since proportional taxes are more useful in relation to reducing toxic pesticide use, they have been selected for all the subsequent scenarios. The level of the tax is set to high, since low and medium taxes have been found to have little impact on the overall reduction of pesticide use, and generate few revenues that can be reinvested. The IPM incentive interventions are selected so as to show how far different pesticide applications are actually affected by the various combinations, and by changes in the level of the policies. As mentioned before, the results of the scenarios are evaluated in terms of their ability to reduce pesticide use, in particular toxic substance use, and their impact on farm household incomes, while also considering policy costs. Also, the results of the scenarios are discussed with local counterparts in order to assess their practicability and applicability.

3.4.5 SWAT-based pesticide use regulation scenarios[5]

The SWAT-based scenarios reduced the baseline application rates of chlorothalonil (45 to 53) and cypermethrin (54 to 62) in a stepwise fashion, to find the necessary reduction target at which the simulated mean daily concentrations were below the pre-determined eco-toxicological thresholds. The NOEC of chlorothalonil was not exceeded in the baseline scenario, and so was not modelled.

Table 28: Number of days per year when simulated daily concentrations exceed NOEC and PNEC thresholds

Scenario (% of baseline application rate)	Days per year above threshold			
	Chlorothalonil		Cypermethrin	
	NOEC	PNEC	NOEC	PNEC
100% (Baseline)	0	4.4	9.4	26.7
50%	0	2.0	5.9	20.7
25%	0	0.1	2.8	14.8
20%	0	0.1	2.5	14.1
18% (PNEC Chlorothalonil)	0	0	–	–
15%	0	0	1.6	12.0
10%	0	0	0.7	9.9
5%	0	0	0.1	5.2
2.5% (NOEC Cypermethrin)	–	–	0	4.2
1%	0	0	0	1.9
0.04% (PNEC Cypermethrin)	–	–	0	0

Table 28 shows the frequency at which daily concentrations of chlorothalonil and cypermethrin exceeded threshold values per year, as simulated by SWAT (Bannwarth et al., 2014). For chlorothalonil, a 50% reduction in use nearly halves the daily events per year in which mean concentrations exceed the PNEC.

5 Parts of this section have been submitted as: Bannwarth, M., Grovermann, C., Schreinemachers, P., Ingwersen, J., Lamers, M., Berger, T., Streck, T., subm. Non-hazardous pesticide concentrations in surface waters: Simulating application thresholds and resulting farm income effects. Journal of Environmental Management.

At 25% of the baseline application rate, the number of days is nearly zero, while at 18%, the PNEC threshold is not exceeded on any day. For cypermethrin, a 50% reduction also nearly halves the number of days at which cypermethrin concentrations exceed the NOEC, but the number of days with concentrations above the PNEC decline only by a fifth. At 10% of the baseline application rate, the PNEC threshold is still exceeded on at least 10 days per year. To avoid exceeding the threshold on any day would require reducing cypermethrin use to 0.04% of the baseline, which basically means that this pesticide cannot be used at all. Table 29 lists the scenarios used in MPMAS to assess the impact of the above mentioned reductions on farm agent incomes.

Table 29: Chlorothalonil and cypermethrin pesticide use reduction scenarios simulated by MPMAS

Scenario ID	Intervention	Percentage reduction of baseline dosage	Scenario names
45–53	Chlorothalonil reduction	50%, 25%, 20%, 18% (PNEC), 15%, 10%, 5%, 2.5% & 1%	CHL500, CHL250, CHL200, CHL180, CHL150, CHL100, CHL050, CHL025, CHL010
54–62	Cypermethrin reduction	50%, 25%, 20%, 15%, 10%, 5%, 2.5% (NOEC), 1% & 0.04% (PNEC)	CYP500, CYP250, CYP200, CYP150, CYP100, CYP050, CYP025, CYP010, CYP004

4. Model verification and validation

Verification and validation is a crucial part of the overall MPMAS modelling process. It ensures that model inputs based on the lottery as well as model outcomes correspond to reality. In fact, validation is considered one of the most important steps in any quantitative modelling approach (Kaiser and Messer, 2012), e.g. farm level decision-making models. A review of 48 bio-economic farm models by Janssen and van Ittersum (2007) however showed that of the reviewed applications only 23 were validated, four of these quantitatively. The MPMAS application used here is based on established decision-making, crop growth and innovation diffusion modelling methods, while the model's coefficients are estimated based on comprehensive and reliable survey data. This approach guarantees the construct validity of the model. In the following section, the verification and validation applied to the model's results are explained in more detail. It should be noted however, that due to the nature of the model and the data collection process, the simulated and observed data are not independent.

4.1 Verification of asset allocations

Verification in the context of MPMAS implies checking that the resources allocated to agents are consistent with the observed resources available to farmers.

In the study model, consistency is tested on six important assets: household size, liquidity, greenhouses owned, area under chrysanthemum cultivation, area under litchi orchards and area under rose cultivation. All of these variables used in the test are expressed as per household quantities. Using linear regression without a constant, the regression line can be predicted. A slope coefficient and an R-squared value close to unity indicate a good fit between the outcome of the asset allocation as calculated using the lottery process, and the asset allocation recorded in the survey. Table 30 shows the goodness of fit parameters for different seed values, those used to initiate the Monte Carlo simulation of the lottery. Seed values affect the random allocation of assets, which requires testing the robustness of different agent population configurations over different seed values. As can be seen from Table 30, the lottery produces robust results. Scenario analysis is also carried out with a seed value of 100. Figure 18 displays the fit between the model and observed data for the selected seed value, with scatter values in the graph representing the outcomes according to clusters owning that particular asset.

Table 30: Goodness of fit and robustness – MPMAS asset allocation

Seed	Household size		Liquidity		Greenhouses		Chrysanthemum		Litchi		Roses	
Value	(member/ hh)		(1000 baht/ ha)		(#/hh)		(ha/hh)		(ha/hh)		(ha/hh)	
	Coef.	R2	Coef.	R2	Coef.	R2	Coef.	R2	Coef.	R2	Coef.	R2
1	0.974	0.989	1.089	0.936	0.908	0.967	1.035	0.994	1.030	0.980	1.017	0.997
3	0.948	0.988	1.013	0.963	0.985	0.868	0.979	0.993	1.015	0.968	0.984	0.996
5	0.930	0.986	1.058	0.965	0.894	0.934	1.100	0.999	1.020	0.966	0.972	0.995
10	0.923	0.991	1.035	0.964	0.907	0.961	0.964	0.993	1.048	0.959	1.019	0.998
20	0.917	0.990	1.048	0.974	0.932	0.961	0.921	0.985	1.045	0.957	1.010	0.998
30	0.981	0.991	1.021	0.979	1.003	0.937	1.070	0.992	1.033	0.967	0.992	0.995
40	0.902	0.990	1.055	0.977	0.943	0.951	0.954	0.995	1.041	0.956	1.018	0.993
50	0.953	0.990	1.062	0.973	0.964	0.953	0.894	0.995	1.033	0.975	0.981	0.997
60	0.983	0.990	1.023	0.956	0.953	0.914	0.909	0.986	1.016	0.976	0.996	0.993
70	0.954	0.990	1.034	0.966	0.908	0.945	1.044	0.991	1.063	0.972	0.992	0.994
80	0.915	0.987	0.925	0.976	0.985	0.941	0.895	0.979	1.015	0.965	1.049	0.994
90	0.951	0.988	1.024	0.955	0.943	0.953	0.819	0.982	1.011	0.973	1.017	0.994
100	0.995	0.990	0.954	0.988	0.925	0.960	1.084	0.999	1.047	0.961	1.072	0.994
150	0.945	0.991	1.050	0.961	0.926	0.925	1.034	0.984	1.012	0.964	1.011	0.993
200	0.981	0.990	1.030	0.976	0.933	0.863	0.949	0.993	1.020	0.970	1.028	0.990
300	0.963	0.990	1.083	0.960	0.880	0.952	1.078	1.000	1.033	0.962	1.045	0.996
500	0.950	0.990	1.023	0.975	0.873	0.943	0.964	0.991	1.026	0.977	1.032	0.992
750	0.953	0.988	0.991	0.989	0.923	0.944	1.071	0.999	1.052	0.960	0.959	0.996
999	0.926	0.990	1.018	0.971	0.955	0.922	0.951	0.956	1.021	0.970	0.962	0.996
Average	0.950	0.990	1.028	0.969	0.934	0.936	0.985	0.990	1.031	0.967	1.008	0.995

4.2 Validation of outcome variables

Validation is predicated on calibrating the model by changing the coefficients in the yield balance rows for the selling activities, and by testing the goodness of fit after each modification. This testing process is carried out for three key outcome variables: land-use, sales revenues and pesticide use. The model runs over five years, and for the validation, results from period two are taken first, since the model is subject to an adjustment processes during the first period. Similar to the verification process, the goodness of fit between reality and the model outcomes is determined by regressing the observed data from the farm household survey

Figure 19: Verification of asset allocation – checking the model fit by plotting observed against simulated values

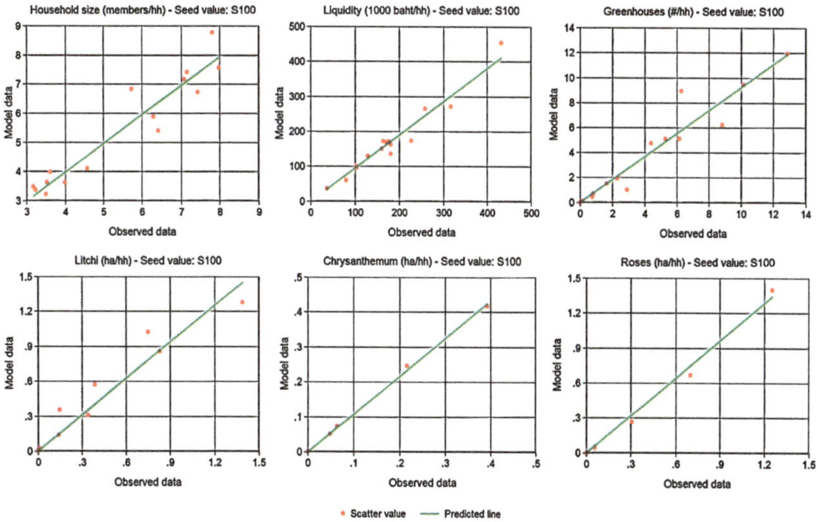

against the simulated data generated by the model (forcing the intercept through the origin). Likewise, the parameters used to evaluate the goodness of fit are the slope coefficient of the regression line and the R-squared. Coefficients between 0.95 and 1.05, and R-squared values between 0.95 and 1, are deemed sufficient for each of the three outcome variables at the aggregate level. This criterion is used in order to determine the suitable final calibration set-up. As can be seen from Table 31, the criteria are met for this application, across the various configurations of the agent population produced by different seed values. The model is thus able to produce a robust representation of land-use in the real world, as well as of the revenues and pesticide use values associated with it.

Table 31: Validation results for three outcome variables across all seed values

Seed	Land-use		Sales revenues		Pesticide use	
Value	Coeff.	R2	Coeff.	R2	Coeff.	R2
1	0.997	0.995	0.953	0.961	0.980	0.980
3	0.987	0.995	0.973	0.967	0.980	0.981
5	0.976	0.994	0.960	0.962	0.979	0.981
10	0.993	0.995	0.955	0.963	0.979	0.981
20	1.004	0.996	0.966	0.963	0.981	0.980

Seed	Land-use		Sales revenues		Pesticide use	
Value	Coeff.	R2	Coeff.	R2	Coeff.	R2
30	0.988	0.995	0.972	0.964	0.980	0.980
40	1.000	0.996	0.958	0.951	0.978	0.982
50	0.985	0.995	0.953	0.960	0.981	0.981
60	0.988	0.996	0.944	0.956	0.979	0.981
70	1.003	0.996	0.949	0.965	0.980	0.981
80	0.982	0.994	0.964	0.964	0.981	0.980
90	0.981	0.995	0.963	0.970	0.981	0.980
100	0.993	0.996	0.969	0.970	0.985	0.981
150	0.982	0.995	0.953	0.958	0.980	0.981
200	0.997	0.996	0.960	0.963	0.979	0.981
300	0.990	0.995	0.962	0.963	0.981	0.981
500	0.995	0.996	0.950	0.96	0.980	0.981
750	1.001	0.997	0.966	0.961	0.981	0.980
999	0.984	0.995	0.965	0.963	0.982	0.980
Average	0.991	0.995	0.960	0.962	0.980	0.981

The scenario analysis is based on a seed value of 100, producing an above average fit for all three outcome variables. For this value, Figure 20 graphically illustrates the validation of the model at the aggregate level. The x- and y-axes show the percentages among the outcome indicators for the observed and the model data respectively. Validation of the first two variables is carried out at the crop level and each dot therefore represents one of the 14 crops. For pesticide use, as shown in the right-hand diagram, the values are aggregated to six WHO toxicity classes. In the first graph, the outlying value is litchi, whereas in the second graph it is greenhouse vegetables, which generates considerable revenues. WHO pesticides classified as unlikely to cause harm produce the outlier in the last graph.

Figure 20: Validation at the aggregate level – checking the model fit by plotting observed versus simulated values

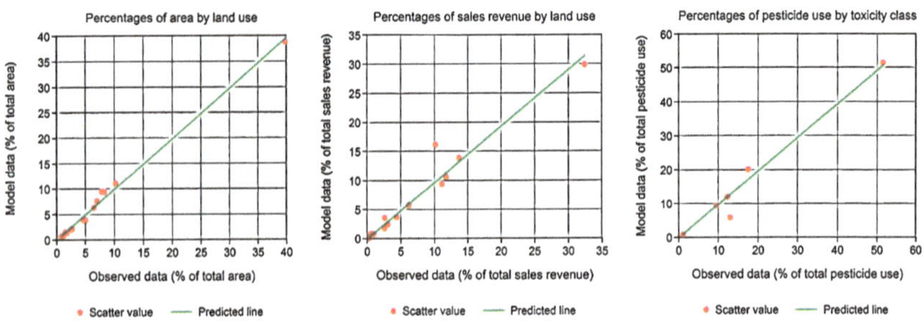

As explained before, the agent population is subdivided into 15 clusters, and this allows for a validation of the model to take place at a less aggregated level. For testing goodness of fit at the cluster level, regressions are run for all three outcome variables in each cluster. Table 32 displays the mean, standard deviation as well as maximum and minimum values of the coefficient of determination and R-squared parameters from these regressions. While the goodness of fit does not match the validation results at the aggregate level, average coefficients ranging from 0.805 to 1.040, with standard deviations between 0.075 and 0.207, can be regarded as acceptable. The same applies to the summary statistics for the R-squared values.

Table 32: Validation at the cluster level – summary statistics of goodness of fit parameters

	Land-use		Sales revenues		Pesticide use	
	Coef.	**R2**	**Coef.**	**R2**	**Coef.**	**R2**
Obs.	15	15	15	15	15	15
Mean	0.851	0.806	1.040	0.878	0.805	0.778
Std. dev.	0.169	0.159	0.075	0.098	0.207	0.213
Min.	0.566	0.514	0.847	0.661	0.453	0.380
Max.	1.114	0.988	1.177	0.997	1.146	0.996

4.3 Testing of innovation diffusion and adoption process

Since IPM for leafy vegetables is an innovation which so far had not been adopted in the study area, validation of the innovation diffusion and adoption process in the model is not straightforward. As described in Section 3.3.7, the innovativeness ranking of agents is based on their knowledge and adoption of GAP certification. To test the accuracy of the innovation diffusion process, it is only possible to compare how far the adoption of IPM corresponds to the adoption of GAP certification. Under these circumstances, an exact substantiation of the results, as carried out for land-use, sales revenues and pesticide use, is not possible. Therefore, for the purpose of testing the modelled innovation diffusion process, two set-ups for the assignment of innovativeness scores (0–4) to agents are evaluated; the random assignment and the direct assignment, and these are compared to a scenario without any network constraints in place. Random assignment implies that the agent innovativeness scores have been produced by a random process, that is the Monte-Carlo simulation of the lottery. The cumulative distribution function for this is segmented according to the adopter category thresholds defined by Rogers (2003), after which agents are randomly assigned to one of the five categories. The direct assignment process refines the allocation of innovativeness scores to agents by using information on the characteristics of the farm households and their networks. These characteristics were previously allocated to agents using the Monte-Carlo method, and then used as explanatory variables. The exact procedure is described in Section 4.2. Table 33 shows the model outcomes for IPM adoption process across the two assignment approaches, as well as the scenario without network constraints plus the observed data for GAP adoption. The results of the direct assignment approach are much closer to the observed data than those of the RAN__________ or the ALL__________ scenarios. Especially in terms of the stage of the innovation diffusion process, the match between the direct assignment of innovativeness and reality is good, since access in both cases is in the first three adopter groups. The different profitability of the growing activities related to innovations might explain why the area under IPM is higher than that under GAP. Generally, it can be said that the direct assignment produces a much more reasonable outcome than the random assignment of innovativeness, or a simulation without network constraints.

Table 33: Comparison of simulated and observed innovation diffusion and adoption

Data	**Innovation**	**Scenario name**	**Access for adopter groups**	**Time since introduction**	**Implementation of the innovation**
Simulated	IPM – access for all segments from period 1	ALL______	No network constraint	5 years (simulation periods)	470.7 ha under IPM
Simulated	IPM – random assignment	RAN______	Innovators only	5 years (simulation periods)	7.5 ha under IPM
Simulated	IPM – direct assignment	IPM_______	Innovators and early adopters and early majority	5 years (simulation periods)	207.2 ha under IPM
Observed	GAP certification		Innovators, early adopters and early majority	6 years (at time of data collection)	79 ha under GAP certification

5. Results

5.1 Private and social levels of optimal pesticide use and overuse[6]

Table 34 shows coefficients for the production functions for the two types of land use. The adjusted R-squared is 0.41 for leafy vegetables and 0.30 for greenhouse vegetables, which is comparable to previous studies in China and Thailand which used the same functional form (Huang et al., 2002; Pemsl et al., 2005; Praneetvatakul et al., 2003).

Table 34: Production function estimates with abatement specification

Variable	Leafy Vegetables		Greenhouse Vegetables	
	Coeff.	SE	Coeff.	SE
Spraying method (1=preventive)	0.301***	3.47	0.295**	2.05
Education (1=low)	-0.102	-1.06	0.054	0.41
Irrigation (1=using)	-0.464***	-4.76		
Location 1 (1=Thai villages at high altitude)[1]	0.176	0.64	-0.270	-1.49
Location 2 (1=Hmong villages at middle altitude)[1]	0.046	0.19	-0.081	-0.43
Location 3 (1=Hmong villages at high altitude)[1]	0.599***	2.89	0.585**	2.13
Labour (baht/ha/month, ln)	0.346***	4.83	0.251***	2.69
Fertilizers (baht/ha/month, ln)	0.130**	2.02	0.355***	3.84
Other inputs (baht/ha/month, ln)	0.065	1.23	0.121*	1.70
Constant	7.086***	12.99	5.422***	5.71
Damage abatement effect (λ) of pesticides (baht/ha/month)	0.0182***	3.28	0.0019**	2.18
N	265		188	
Adj. R^2	0.41		0.30	

Notes: Dependent variable is output in ln(baht/ha/month). Omitted location dummy is Thai villages at middle altitude.

The crop dummies are not shown. Significance levels: *P < 0.10, **P < 0.05, ***P < 0.01.

6 This section has been published as part of the following article: Grovermann, C., Schreinemachers, P., Berger, T., 2013. Quantifying pesticide overuse from farmer and societal points of view: An application to Thailand. Crop Protection 53, 161–168.

The dummy variable identifying whether farmers' spraying is mainly preventive as opposed to responsive, has a positive and significant effect on output, while the effect of having a low education is insignificant. The use of irrigation has a significant negative effect on output, which suggests an intervening effect of seasonality or management, as yields in the dry period are generally lower. The effect of all growth-stimulating inputs is positive and significant, with labour having the highest coefficient.

The regression coefficient for pesticides is positive and significant for both land uses. Figure 21 shows the functional shape of the damage abatement term, revealing that the abatement effect reaches 100% and levels off at a relatively small quantity of pesticide. The shape is similar for all land use groups, but the x-scale is different as farmers use much higher amounts of pesticides on greenhouse vegetables.

The marginal value product of pesticides is estimated for each farmer in the sample, with the marginal value product being the value of output resulting from one additional Baht spent on pesticides. Values above unity hence point at the underuse of pesticides from a private point of view while values below unity point at overuse. The marginal value product is below unity for 86.1% of the observations for leafy vegetables, and 88.7% of the observations for greenhouse vegetables. While up to the 75th percentile the marginal value product is close to zero, the average marginal value product is greater, at 0.39 and 0.56 for leafy and greenhouse vegetables respectively, due to a few outlying data points sitting well above unity.

Figure 21: Effects of pesticide use on crop yields for leafy and greenhouse vegetables

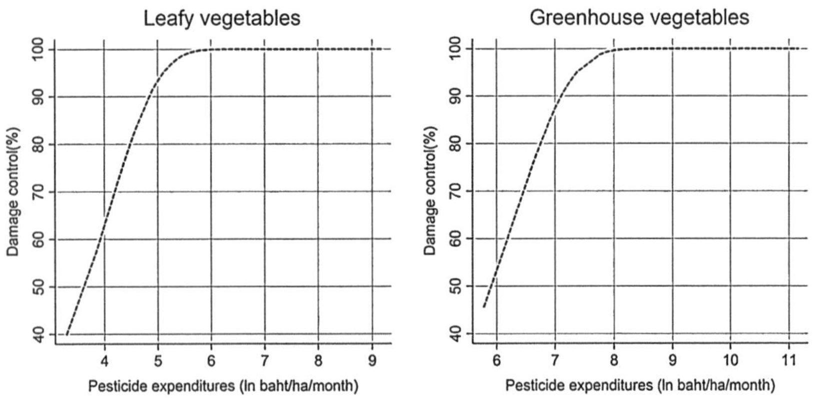

Optimal levels of pesticide use are also determined for each observation, depending on the costs associated with the applied pesticides as well as the use of other inputs. Table 35 shows that the average optimal levels of pesticide use are relatively small and the levels of overuse relatively high. This applied to both private and social levels of overuse, the difference between the two being relatively small though. As a consequence of the consistently low marginal productivity shown for the majority of pesticide use observations, a very substantial amount of pesticides, 79% for leafy vegetables and 78% for greenhouse vegetables could thus be categorized as overuse from a private point of view. Because of the relative steepness of the exponential damage control function, as shown in Figure 20, adding external costs to the private costs only has a minor effect on the quantity of overuse.

Table 35: Private and social levels of optimal pesticide use and overuse

	Leafy Vegetables		**Greenhouse Vegetables**	
	Private	**Social**	**Private**	**Social**
Av. optimal use (1000 baht/ha/month)	0.34	0.32	2.89	2.77
Total overuse (1000 baht)	336	340	1,287	1,302
Overuse (as % of total quantity)	79	80	78	79

Notes: Overuse determined for each observation as the difference between actual and the private/social optimal pesticide use. Total overuse for the whole watershed is the sum of these individual differences.

5.2 Adoption of GAP standard

The probit model with sample selection used to estimate the adoption probabilities based on the survey farm-level data resulted in a Wald chi2 of 342.77 (Prob > chi2 = 0.000). Therefore, as a whole, the model is statistically significant. At the same time, the likelihood-ratio (LR) test indicates that the results of fitting the combined model, which corrects for the sample selection bias, are significantly different from the outcomes produced when separately fitting the selection model for the knowledge of GAP and the probit model for GAP certification (chi2 = 3.88; Prob.> chi2 = 0.049). It is thus necessary to use the full model instead of a simple probit regression.

Table 36 shows the regression output. As far as the estimated coefficients are concerned, it should be noted that they represent the effect of each explanatory variable on the log of the odds of an outcome occurring {ln[P/(1-P)]}, rather than the effect of a change in the variable on the probability of an outcome (Neupane et al., 2002). Even though some of the variables might be conceptually related, the variance inflation factor used was 1.99, which suggests that multi-collinearity is not an issue.

Table 36: Probit regression with sample selection – Output

	GAP standard certification (probit model)		GAP knowledge (selection model)	
Variables	**Coef.**	**SE**	**Coef.**	**SE**
Household size (#)	0.023	0.051	0.007	0.042
Farm size (ha)	−0.386***	0.103	−0.093	0.081
Percentage of high value vegetables (%)	0.860**	0.399	0.315	0.251
Percentage of litchi (%)	2.023***	0.418	0.363	0.349
Farm age (years)	−0.045**	0.020	0.014	0.012
Age household head (years)	0.070***	0.024	−0.006	0.012
Education (yes=1)	1.148***	0.291	0.809***	0.223
Grow more than 1 crop (yes=1)	0.821**	0.396	0.339	0.214
Applying pesticides regularly (yes=1)	0.361	0.223	0.345**	0.174
Liquidity per capita (1000 baht)	0.003***	0.001	0.001	0.001
Own motorbike (yes=1)	0.668*	0.344	0.379*	0.237
Member of Royal Project (yes=1)	1.502***	0.229	1.054***	0.163
Exposure (#)	2.201***	0.607	2.194***	0.439
Village head (yes=1)	−1.009*	0.543	−0.021	0.454
Born in the Mae Sa watershed (yes=1)			−0.014	0.168
Constant	−7.506	1.072	−3.163	0.669
N	111		295	
Wald chi2 = 342.778, Prob. > chi2 = 0.000				
LR test of independent equations (rho = 0): Chi2 = 3.88; Prob. > chi2 = 0.049				

Note: Significance levels: *P < 0.10, **P < 0.05, ***P < 0.01

In the selection model, 15 variables are used, which is one more than in the probit model. This is done in order to guarantee that both models are structurally different. Education, membership of the Royal Project extension organisation, as well as exposure to peers who know about an innovation, have a highly significant, positive effect on the level of knowledge of the GAP standard. Also, farmers who regularly apply pesticides and own a motorbike, so increasing their mobility, are more likely to know about the GAP standard. Education, Royal Project membership, exposure and owning a motorbike are significant in the same positive direction in the probit model as in the selection model. The percentage of high value vegetables as well as litchi on the farm, which are included as control variables, the age of the household head, diversification (growing more than one crop) and liquidity per capita in the farm household, are other positive and significant determinants of certification with the GAP standard. In contrast, farm size, farm age and having a village head in the household have a significant negative impact on the adoption of the GAP standard. Bigger and older farm households, or those with a household member in a traditional leadership position are therefore likely to be less innovative.

Table 37: Marginal effects of the probit model within the sample selection

Variable	dy/dx	SE
Household size (#)	0.004	0.009
Farm size (ha)	–0.063***	0.020
Percentage of high-value vegetables (%)	0.141**	0.069
Percentage of litchi (%)	0.332***	0.092
Farm age (years)	–0.007**	0.004
Age household head (years)	0.012**	0.005
Education (yes=1)	0.122***	0.034
Grow more than 1 crop (yes=1)	0.109***	0.039
Applying pesticides regularly (yes=1)	0.062	0.042
Liquidity per capita (1000 baht)	0.001**	0.000
Own motorbike (yes=1)	0.084**	0.035
Member of Royal Project (yes=1)	0.329***	0.065
Exposure (#)	0.362***	0.104
Village head (yes=1)	–0.088***	0.028

Note: Significance levels: *P < 0.10, **P < 0.05, ***P < 0.01

The marginal effects in Table 37 show an above 30% increase in the probability of adoption with changes in the three independent variables: The proportion of litchi cultivation, Royal Project membership and exposure all have a strong effect. Farmers growing litchi are interested in GAP certification since it allows them to access more lucrative markets (Schreinemachers et al., 2012). The same is true for farmers growing high-value vegetables, where the marginal effect is 14%. Positive changes in the adoption probability can also be observed when a household owns a motorbike, among older household heads and when there is higher liquidity per capita in the household, though for the latter the marginal effect is very small. If a member of a household is village head, then the probability of adoption drops by 8%. The marginal effects for farm size and farm age are also negative. Household size and the regular application of pesticides have no significant effects, even though the coefficient of the second variable turned out to be significant in the selection model. The prediction that can be derived from the econometric model gives insights into the relative importance of household and farm characteristics in terms of adoption, and how adoption rates could be increased; for example by boosting membership of the Royal Project or improving communications within the innovation network. As Schreinemachers et al. (2009) point out, as a comparative static model that lacks system dynamics, the econometric model cannot predict when adoption will occur, so is useful when used alongside and complementing an agent-based simulation.

Using the coefficients presented in Table 36, which are estimated from the sample, it is possible to predict the adoption probabilities for the agent population with the explanatory variables allocated to each agent by the lottery (for details see sections 4.2.2, 4.2.3 and 4.3.7). Agents can be ranked and divided into the five adopter categories. None of the groups uses IPM for leafy vegetables, so that only innovators have access initially and the next groups need to wait until the network threshold is reached, which corresponds is the upper bound of the percentage of adopters in each group, this being 2.5% for the first threshold. As Figure 22 shows, the average probability of adoption clearly declines from the first adopter group, through the innovators and on to the fifth group – the laggards.

Figure 22: Predicted adoption probabilities in the five adopter categories

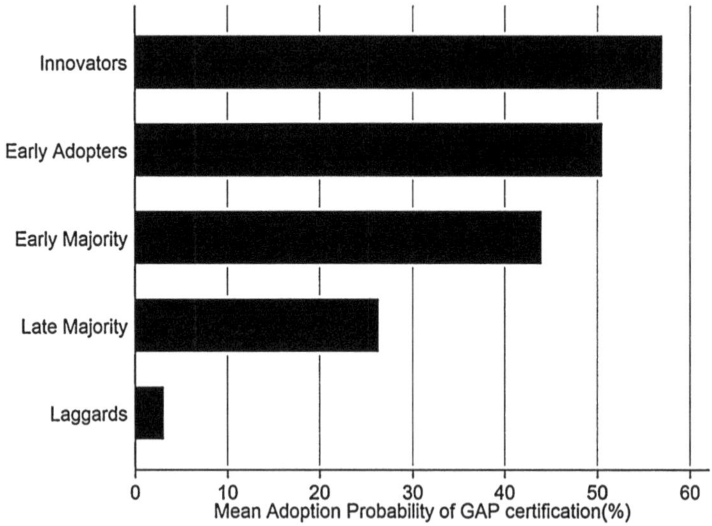

5.3 Simulation experiments

The scenario analysis conducted for this study includes simulations on the impact of different taxation schemes, both as stand-alone interventions and in combination with the reimbursement of tax access to IPM and other policy interventions. These interventions involve incentives such as subsidies for bio-pesticides, price premiums for IPM produce and area payments for agricultural land under IPM. In the following section, the results of the different scenarios are presented roughly in order of increasing complexity as regards the mix of policy interventions. Moreover, the scenarios designed with the pesticide concentration thresholds obtained by the SWAT model are explained at the end of the chapter. Altogether, the following sections provide an overview of the impacts of a wide range of economic instruments and science-based regulations, those that aim to reduce pesticide use. The performance of these interventions is assessed in terms of the reductions made in overall and/or toxic pesticide use, of the changes farm agent incomes and costs and, if applicable, the diffusion of IPM. In this study, toxic pesticide use refers to the application of especially hazardous pesticides, those that belong to the WHO I and WHO II toxicity classes.

5.3.1 The baseline scenario

The baseline scenario serves as the benchmark for all further analysis. The outcomes of all simulation experiments that follow are compared in terms of pesticide use reduction as well as income and land-use changes against the baseline scenario. Figure 23 shows the cumulative distribution of pesticide use and income as against the cumulative distribution of agents in the first and last time periods of the baseline scenario. The straight diagonal line indicates that the proportion of pesticide use or income is equal to the proportion of agents, which is not the case in the simulated results here. The red lines below the diagonal line provide evidence of an unequal distribution of both variables in the agent population. In this regard, there is no difference between period 1 and period 5. While 60% of agents are responsible for less than 30% of total applied pesticides, they also earn less than 25% of the total cash earnings in the watershed. This is similar to estimations made from the survey data (Schreinemachers et al., 2011), but the disparity between farmers heavily applying pesticides and those using only a few is even more drastic for the survey data than in the model.

Income inequality and disproportionate pesticide use are an important justification for using an agent-based modelling approach, and Figure 23 shows that both are related. On the left side of the figure, the joint distribution of pesticide use

Figure 23: Cumulative distributions of pesticide use and income in the first and last time period

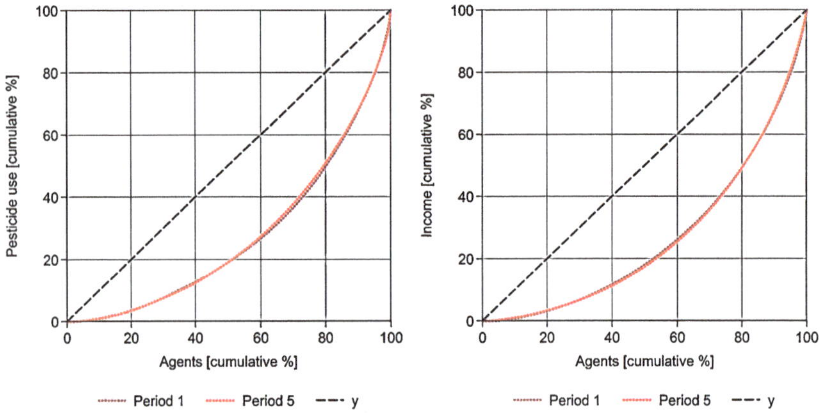

and income is shown for the first and the last time period. Over time, a clear consolidation becomes apparent; incomes and pesticide use among farm agents increase somewhat more in the lower ranges, and the spread of data is reduced. The right hand graph represents the input-outcome relationship, pointing out which of the observations are related to the use of more toxic pesticides, defined as the use of more than 50% of the WHO I and WHO II pesticide categories in relation to the total pesticide quantity used. The increased use of toxic pesticides is clearly associated with higher incomes, with the increase in relation to income being linear. For the remaining observations, the relationship is not so clear, especially for higher application amounts, when incomes do not substantially increase, and even decrease in some cases. This is in line with the findings of the econometric pesticide productivity analysis. However, that analysis was carried out at the crop level, and the relationship between pesticide use and farm incomes is not exactly the same as that between pesticide use and output. There is also no clear directional relationship, as overall wealthier households have the means to invest in greenhouses and can spend more on inputs, thus obtaining higher incomes on average.

In recent years, farm households in the watershed have increased their area under profitable and pesticide-intensive greenhouse vegetables and flowers, while at the same time cutting down their litchi trees. This is confirmed when comparing the results of the latest survey with those of a previous survey conducted in the same area in 2006 (Schreinemachers, 2008). The trend for bell peppers and tomatoes, which are grown in permanent greenhouses with steel frames and plastic covers, is reflected by the simulation results, as can be seen in

Figure 24: Relationship between incomes and pesticide use

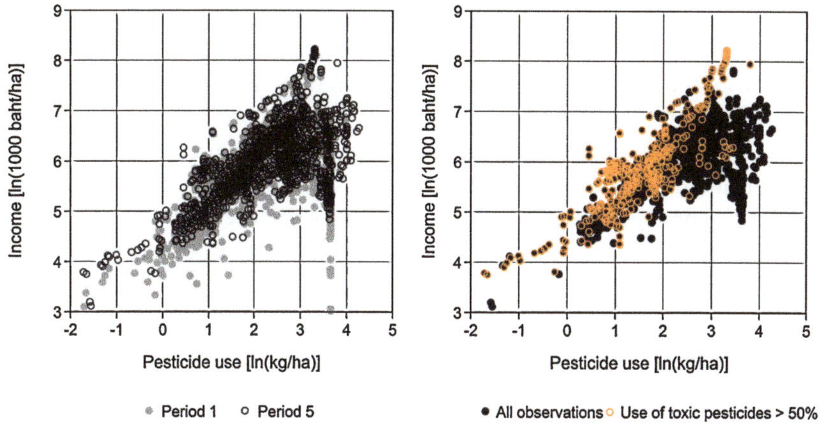

Table 38. Likewise for flowers, simulated land-use grows slightly more important over time. In this context, pesticide use and income can obviously be expected to increase. As well as land-use trends, Table 38 shows the simulated trends for pesticide use in kg per hectare, and the incomes per farm agent household. Averaged over all simulation periods, the total annual income of all 1491 agents amounts to 368.4 million Baht, and pesticide use over a total cultivated area of ca. 1100 ha is 15,497 kg. Simulated pesticide use over time is in the same order of magnitude as that estimated from the survey (Schreinemachers et al., 2011). Since litchi is implemented as a perennial crop in the model, and the area under litchi cultivation is initially allocated to agents, the fruit tree area does not significantly change over time, which can be considered realistic in the short run.

Table 38: Land use, pesticide use and incomes over time

Period	Cereals (ha)	Leafy veg. (ha)	Greenh. veg. (ha)	Other veg. (ha)	Flowers (ha)	Fruit tree (ha)	Pesticide use (kg/ha)	Income (1000 baht/hh)
1	81	246	85	67	181	425	13.41	218.85
2	79	225	103	78	194	419	13.9	238.90
3	77	218	117	78	201	425	14.02	253.43
4	75	212	124	78	202	423	14.14	258.46
5	79	211	131	81	195	425	14.75	265.77

Notes: Results averaged over all agents, total cultivated area in the model: 1100 ha

5.3.2 Impact of tax interventions

The following nine scenarios were developed in order to assess the potential effect of different taxation schemes on the use of chemical pesticides among upland farmers, and to evaluate the related impacts on their livelihoods. The results of each simulation experiment need to be seen in the context of the baseline, as well as the other scenarios. Several graphs and tables are shown here to facilitate comparison, and these also serve in the evaluation of scenarios found in subsequent chapters. As described in Section 4.2, three levels of flat taxes [___TF1_________, ___TF2_________, ___TF3_________] are compared to three levels of proportional taxes [___TP1_________, ___TP2_________, ___TP3_________]. At the same time, a compensation scheme to redistribute the tax revenues back to the farm agents is to assess the added value gained from paying the farmers back their taxes. This results in three additional scenarios [___TC1_________, ___TC2_________, ___TC3_________], in which the proportional tax revenues are reimbursed in the form of a lump sum payment in relation to farm size.

Table 39: Simulated changes in pesticide use and income levels for the different tax scenarios

Scenarios Name	Description	Change in pesticide use (kg)	(kg/ha)	(%)	Change in income (1000 baht)	(1000 baht/ ha)	(%)
___TF1_________	Low flat tax	-413	-0.37	-2.62	-6,114	-5.55	-1.59
___TP1_________	Low prop. tax	-325	-0.29	-2.06	-4,733	-4.28	-1.24
___TC1_________	Low prop. tax + compensation	-300	-0.27	-1.90	-3,290	-2.97	-0.86
___TF2_________	Med. flat tax	-935	-0.86	-5.93	-19,511	-17.86	-5.12
___TP2_________	Med. prop. tax	-794	-0.72	-5.03	-15,066	-13.68	-3.96
___TC2_________	Med. prop. tax + compensation	-809	-0.74	-5.13	-14,273	-12.96	-3.76
___TF3_________	High flat tax	-1,338	-1.22	-8.48	-28,410	-25.95	-7.45
___TP3_________	High prop. tax	-1,157	-1.05	-7.34	-24,910	-22.62	-6.56
___TC3_________	High prop. tax + compensation	-1,179	-1.07	-7.47	-18,308	-16.62	-4.81

Note: Averages over all agents and simulation periods, values represent the difference between the respective scenario and the baseline

Across the range of tax interventions presented in Table 39, pesticide use reductions remain below 10% of the baseline. Flat taxes achieve the greatest pesticide use reductions, but also most severely affect income levels. With the high flat tax rate, the reduction lies at 8.48%, which involves an income loss of 7.45%. The respective numbers for the high proportional tax are 7.34% and 6.56%. While a higher reduction of overall pesticide use is attained with flat tax rates, it is important to note that the reduction effect for the more toxic pesticides is higher when using proportional taxes, as these are targeted. This is illustrated in Figure 25, in which the change in WHO I pesticide use (left) is noticeably bigger for proportional taxes, while differences in the changes to WHOII pesticide use (right) are observable, but less clear.

Figure 25: Change in toxic pesticide use for different tax scenarios

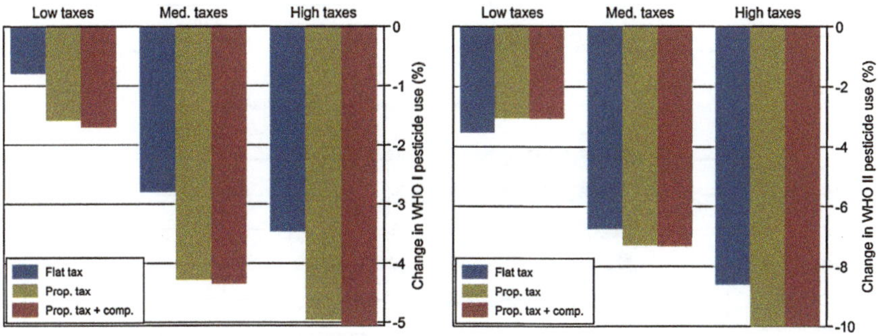

Note: Averages over all agents and simulation periods, bars represent the difference between the respective scenario and the baseline

In addition to the differences between uniform and graduated taxation, it is important to consider differences between the proportional taxes with and without compensation payments. In this regard, it becomes apparent from the simulated outcomes across the three tax levels that changes in pesticide use are negligible, and income losses only partially offset. The changes in land-use induced by the tax levies prevent the baseline income levels being reached; therefore, lump sum payments only compensate for lost earnings to a minor or moderate degree.

While the results presented so far refer to average impacts, the box plots in Figure 26 convey important messages about the effects of the tax interventions on the entire population of farm agents. First of all, across all scenarios, the reductions in pesticide use and the income losses are more pronounced at the higher percentiles of the distribution, and less distinct at the lower percentiles. In the figure, this can be clearly seen from the outer edges of the boxes: the 25th and 75th percentiles. The farm agents spraying fewer pesticides reduce their application amounts to a lesser degree. As was shown above, these agents are also relatively

Figure 26: Distribution of average pesticide use and incomes among the agent population for different tax scenarios

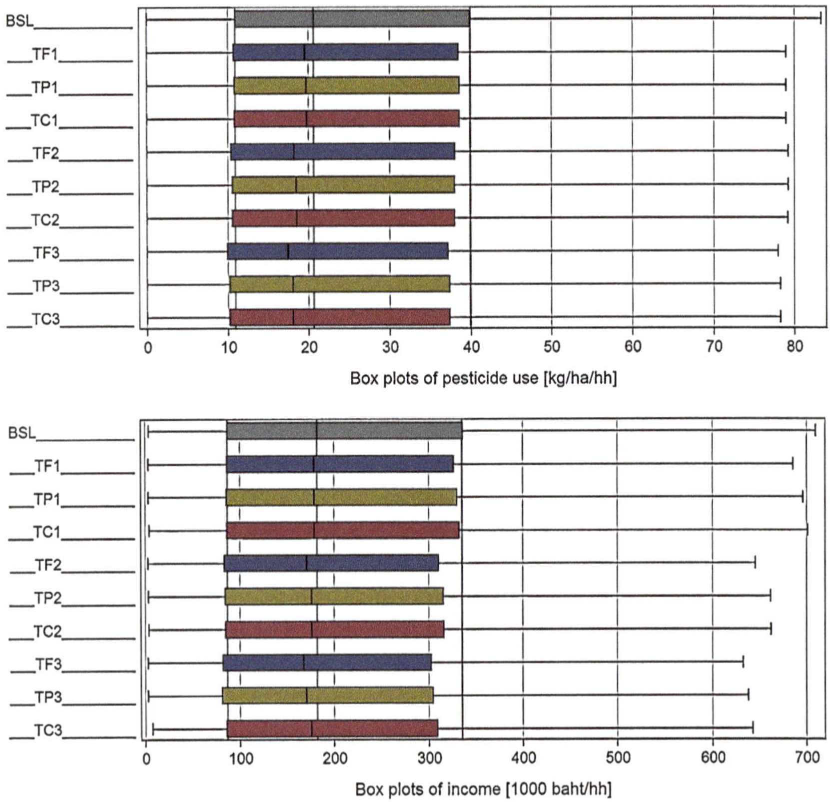

poor when compared to those spraying more excessively, but also less severely affected by income losses. At the median income level and above, the distributions are shifted more distinctly, in particular by the medium and high level tax interventions. Moreover, the box plots demonstrate that the effect of the compensation payment is not substantial, but that poorer households benefit more from it, which is a policy-relevant finding. At the 25th percentile, the reimbursement of taxes almost allows baseline income levels to be maintained. Figure 27 highlights this point, as it shows the cumulative distribution of income. In this figure, from 30% of cumulative agents onwards, the gap between the tax intervention curves with and without compensation starts to narrow. In the lower part of the distribution, the relatively poorer farm agents are basically fully compensated for their income losses.

Figure 27: Cumulative income distribution for high prop. tax with & w/o compensation

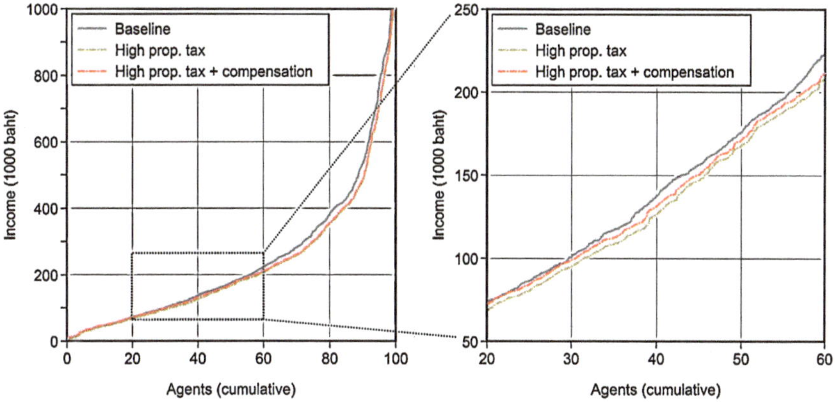

Besides pesticide application amounts and farm household incomes, land-use in the watershed is changed when giving farm agents a disincentive to use chemical plant protection. Crops involving less pollution, such as cereals or chayote (under Other veg.) become comparatively more attractive, but crops involving more pollution, such as onions (also under Other veg.) or bell peppers (under Greenh. veg.) become less attractive.

Table 40: Simulated changes in land-use for the different tax scenarios

Scenario Name	Cereals (%)	Leafy veg. (%)	Greenh. veg. (%)	Other veg. (%)	Flowers (%)	Fruit tree (%)
___TF1________	0.00	0.65	-2.13	-0.37	-0.06	0.16
___TP1________	-0.32	1.41	-7.64	0.55	-2.27	-0.80
___TC1________	-1.38	1.20	-12.67	1.30	-2.43	-1.62
___TF2________	0.94	0.71	-1.11	-1.10	0.43	0.29
___TP2________	0.94	0.71	-1.12	-1.10	0.14	0.29
___TC2________	-0.01	1.97	-6.88	-2.12	-1.19	-0.83
___TF3________	0.00	1.49	-7.08	-2.32	-1.71	-1.25
___TP3________	1.87	2.04	-11.27	-0.19	-0.34	0.22
___TC3________	2.30	2.48	-11.30	0.13	-1.18	0.29

Notes: Results of period 5 averaged over all agents, values represent the difference between the respective scenario and the baseline, total cultivated area in the model: 1100 ha

Under the simulations, the area growing greenhouse vegetables decreases by approximately 1% to 2% for the low tax interventions and goes down by ca. 11% to 12% for the high tax levels, which is a clear contraction path. The decrease in area under flowers is less pronounced. In contrast to these trends, the area under leafy vegetables expands to a lesser but noticeable extent. Agents seem to replace some of their highly polluting bell pepper and tomato production activities with slightly less profitable but also less polluting leafy vegetable growing activities. However, most leafy vegetable activities still require rather substantial amounts of pesticides to be used, which explains why the overall reduction in pesticide use is not more pronounced.

Ultimately, it is crucial to evaluate which policy is the most successful in bringing down levels of pesticide applications. Therefore, it is necessary to analyse outcomes that are expressed in different units, which prevents the use of cost-benefit analysis. For the purpose of the present tax appraisal, we thus analyse effectiveness, which is assessed on the one hand in terms of income loss per percentage reduction of toxic pesticide use (income loss effectiveness) and, on the other hand, in terms of reduced percentage of toxic pesticide use per unit of tax revenue (tax effectiveness). At each taxation level, the flat taxes turn out to be less effective than the gradual taxes, at least when taking into account toxic pesticide use. Across the taxation levels, effectiveness declines from the low to the medium and high taxation levels. The income loss per percentage cut in pesticide use increases slightly, while the percentage reduction per 1000 Baht of tax revenues per agent household falls.

Figure 28: The decision space for tax scenarios

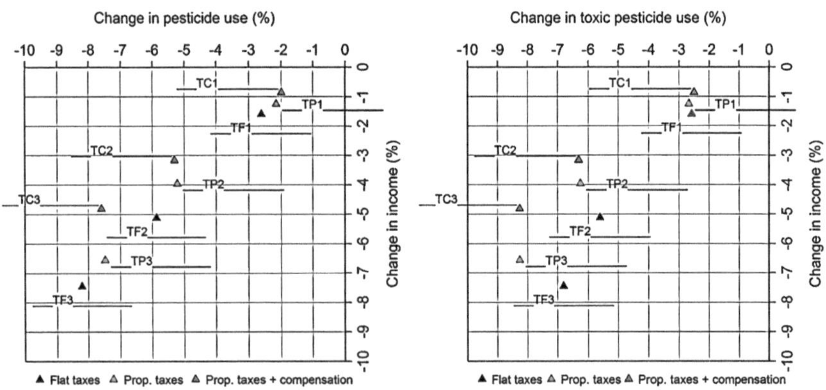

Table 41: Evaluation of interventions for tax scenario impacts when compared to the baseline

Scenario name	Δ income		Tax revenues	Policy costs	Net benefit	Δ pesticide use	Δ toxic pesticide use	Income loss effectiveness	Tax effectiveness
	(%)	(1000 baht/hh)	(1000 baht/hh)	(1000 baht/hh)	(1000 baht/hh)	(%)	(%)	(Δ income/ Δ toxic pesticide use)	(Δ toxic pesticide use/ tax revenues)
TF1	-1.59	-4.10	1.44	0.00	-2.66	-2.62	-2.58	-1.59	-1.79
TP1	-1.24	-3.17	1.48	0.00	-1.69	-2.06	-2.67	-1.19	-1.80
TC1	-0.86	-2.21	1.48	0.00	-0.72	-1.90	-2.50	-0.88	-1.68
TF2	-5.12	-13.09	4.10	0.00	-8.99	-5.93	-5.61	-2.33	-1.37
TP2	-3.96	-10.10	4.21	0.00	-5.89	-5.03	-6.26	-1.61	-1.49
TC2	-3.16	-8.57	4.21	0.00	-5.37	-5.13	-6.32	-1.52	-1.50
TF3	-7.45	-19.05	6.56	0.00	-12.49	-8.48	-6.81	-2.80	-1.04
TP3	-6.56	-16.71	6.74	0.00	-9.97	-7.34	-8.26	-2.02	-1.23
TC3	-4.81	-12.28	6.73	0.00	-5.55	-7.47	-8.27	-1.49	-1.23

Note: Averages over all agents and simulation periods, values represent the difference between the respective scenario and the baseline

The scatter plots in Figure 28 show that the compensation payment is actually only effective for the high proportional taxation scenario. It has a clear impact on abating income losses. Since less pesticide intensive but highly profitable crops are grown, the compensation payment does not mitigate income losses to the extent that baseline levels are reached. For the two variables in the decision space, Figure 27 also points toward a rather linear relationship for different taxation levels.

5.3.3 Impact of IPM adoption with and without pesticide taxes

This section highlights the impact on farm agents of providing access to integrated pest management practices for leafy vegetables. For this purpose, access to the technology is granted to the innovator segment in period 1. As explained previously, the comprehensive IPM methods used for this research are rarely adopted in northern Thailand as a whole, and can be considered a new innovation in the Mae Sa watershed. The tax interventions assessed in the previous section now need to be considered in combination with the introduction of IPM. Therefore, in the following section, the IPM reference scenario without any further changes [IPM____________] is compared to the three levels of flat taxes [IPMTF1_________, IPMTF2_________, IPMTF3_________], to three proportional taxes without lump sum payments [IPMTP1_________, IPMTP2_________, IPMTP3_________] and to three proportional taxes with a compensation payment in place [IPMTC1_________, IPMTC2_________, IPMTC3_________].

Figure 29: Innovation diffusion for IPM + tax interventions

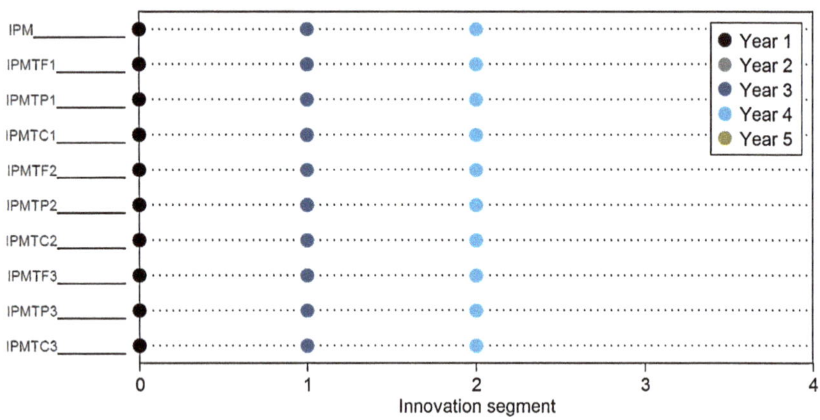

As illustrated by the data presented in Section 2.5, IPM is comparatively profitable on average and should therefore be attractive to profit-maximising agents. The diffusion pathways for each scenario shown in Figure 29 substantiate this. The dots in the figure are coloured according to the year in which the network threshold of the respective innovation segments is reached. The segments represent the five adopter categories (innovators, early adopters, early majority, late majority and laggards). The innovation diffuses rather quickly, so that in year 4, agents in the early majority segment can already adopt IPM. However, there is no difference in diffusion between the scenarios. Even high taxes do not speed up the diffusion process; therefore, cheap pesticides do not appear to constrain the adoption of IPM in the current application.

Table 42: Simulated changes in pesticide use and income for IPM + tax interventions

Scenarios Name	Description	Change in pesticide use (kg)	(kg/ha)	(%)	Change in income (1000 baht)	(1000 baht/ha)	(%)
IPM	IPM access only	–882	–0.73	–5.53	42,216	35.25	10.93
IPMTF1	IPM + low flat tax	–1,214	–1.01	–7.65	37,256	30.89	9.61
IPMTP1	IPM + low prop. tax	–1,138	–0.95	–7.16	39,628	32.92	10.24
IPMTC1	IPM + low prop. tax + compensation	–1,116	–0.93	–7.02	39,272	32.69	10.15
IPMTF2	IPM + med. flat tax	–1,710	–1.45	–10.79	25,504	21.04	6.51
IPMTP2	IPM + med. prop. tax	–1,658	–1.40	–10.46	27,179	22.41	6.97
IPMTC2	IPM + med. prop. tax + compensation	–1,658	–1.40	–10.46	28,235	23.30	7.23
IPMTF3	IPM + high flat tax	–2,152	–1.82	–13.58	17,220	14.03	4.34
IPMTP3	IPM + high prop. tax	–2,042	–1.73	–12.90	19,257	15.52	4.84
IPMTC3	IPM + high prop. tax + compensation	–2,030	–1.72	–12.83	23,852	19.35	6.04

Note: Values are averages over all agents and simulation periods, values represent the difference between the respective scenario and the baseline

While differences in diffusion are not apparent, there is a clear impact on the two main outcome variables evaluated in this study: pesticide use and income. The adoption of IPM leads to a situation in which agents are on the whole considerably better off. As Table 42 shows, making IPM available to agents increases their overall income level by almost 11%. The increase is lower, but still significant, once taxes are introduced, ca. 5% for a high proportional tax without and 6% for that with compensation payment. Whereas pesticide use reduction remains below 10% for all tax interventions simulated without IPM access, the introduction of IPM pushes the reduction across the 10% milestone for all medium and high tax interventions, and up to 13.6% in the case of high flat taxes. IPM access without any further intervention [IPM____________] reduces pesticide use by 5.5%, which means that the reduction effect of the innovation and the taxes is not strictly, but roughly additive. This can be seen by comparing the results presented in Table 39 with those from Table 42.

Figure 30: Distribution of pesticide use and incomes in the agent population for IPM + tax interventions

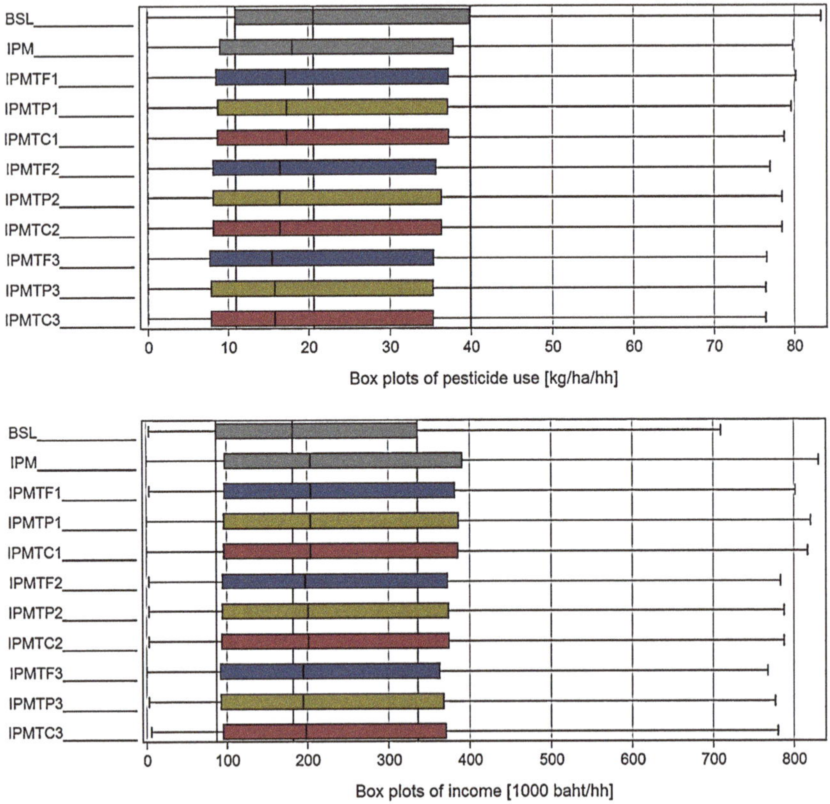

Contrary to the findings presented in the previous section when examining the agent population as a whole, the reduction in applied pesticides is similar or even slightly more prominent at the 25th percentile when compared to the 75th percentile of observations, due to the impact of IPM access. In relative terms, the reduction is considerably greater at lower levels of pesticide use, while overall it is most marked at the median level. The very heavy use of synthetic pesticides is not affected much by the interventions, as the outer whiskers of the box plots show. The picture is different when investigating the income effects, for here it becomes clear that the distribution shifts more clearly towards the right at the higher percentiles. At the 25th and 50th percentile, the income gains are more moderate than at the 75th percentile. The adoption of IPM in the agent population is fairly evenly distributed, as Figure 31 illustrates. On the right hand side in the figure, the plots reveal pesticide use and incomes for all agents in the baseline scenario [BSL____________], with IPM adopters under the IPM introduction in combination with high proportional taxes scenario [IPMTP3_________] marked in orange and red. In that scenario, of all 1491 agents, 654 adopt IPM practices by period 5. While the mean agent household income level of adopters rises from the baseline income of 260,000 to 288,000 Baht under IPM in combination with high proportional taxes [IPMTP3_________], the mean income of non-adopters drops from 270,000 Baht to 251,000 Baht in the respective scenarios. The plot on the right hand side in Figure 31 shows pesticide use and income levels for all agents, highlighting IPM adopters in the scenario in which IPM is introduced alongside taxes [IPMTP3_________]. The two diagrams in Figure 30 illustrate that the income effect observed in the box-plot is thus due to a partial redistribution of incomes among the agent population. The relationship between pesticide use and incomes is weakened; however, as IPM adopters across the income spectrum seem to benefit.

Figure 31: Income levels and pesticide use, highlighting IPM adopters in scenarios BSL & IPMTP3

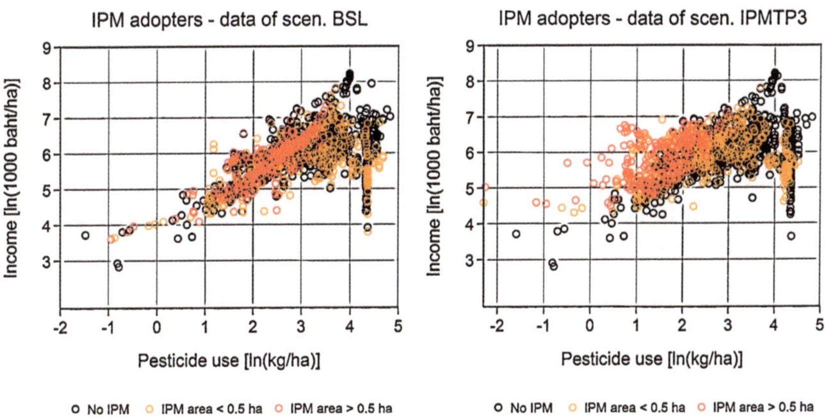

It is of interest to examine more closely how far agents gain or lose from the introduction of IPM in combination with taxes [IPMTP3__________], and so Figure 32 helps to understand what determines changes in income with regard to the initial income situation, the area under IPM and the innovativeness of the individual agents in this scenario. The figure shows the 15 clusters into which the agent population is subdivided (for more explanation see the Chapter 3). The first graph on the left clearly identifies that gains and losses occur across the lower and middle ranges of the average baseline income level, with gains clearly outweighing the losses in terms of magnitude. The graph in the middle shows that income declines for those clusters in which agents have not adopted IPM. The graph on the right then reveals that the innovativeness of agents in the clusters is a major driver of changes in income. The less innovative clusters, with average innovativeness scores above 2.5, are without exception those in which income losses are registered. Therefore, innovativeness can be seen as a major factor determining who benefits from the introduction of IPM.

Figure 32: Income change related to baseline income, IPM area & innovativeness (15 clusters, BSL vs. IPMTP3)

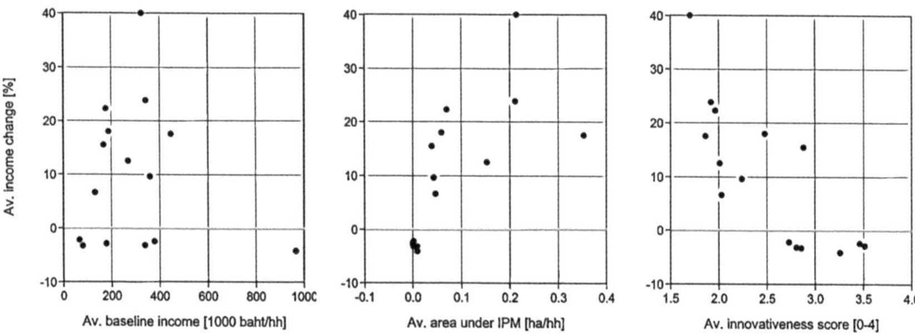

As expected from the diffusion of IPM, the area under IPM is not appreciably different under the ten scenarios; however, the overall land-use change brought about by the introduction of IPM practices in the watershed across the scenarios is substantial. In each of the scenarios, the area under IPM in the last simulation period amounts to almost 20% of the total cultivated area. The increase of this land-use type occurs mainly at the expense of the area under onions, beans and chayote (Other veg.) and under bell peppers, as well as tomatoes (Greenh. veg.).

Table 43: Land-use changes – IPM + tax interventions

Scenario Name	Cereals (%)	Leafy veg. (%)	Greenh. veg. (%)	Other veg. (%)	Flowers (%)	Fruit tree (%)	IPM (ha)
IPM____________	-0.81	-19.31	-12.36	-34.71	-2.37	-1.17	207.25
IPMTF1________	-0.80	-18.78	-14.59	-35.42	-3.02	-0.46	210.32
IPMTP1________	0.26	-18.81	-13.82	-36.57	-2.99	-0.91	210.45
IPMTC1________	0.26	-18.74	-13.55	-36.57	-2.34	-0.42	210.99
IPMTF2________	-1.23	-18.63	-20.01	-35.14	-4.62	-1.99	213.75
IPMTP2________	-1.02	-18.49	-19.93	-38.08	-4.35	-0.93	213.39
IPMTC2________	-0.73	-18.15	-19.40	-38.03	-4.36	-1.32	215.06
IPMTF3________	-1.94	-18.58	-26.15	-34.30	-5.38	-1.85	213.85
IPMTP3________	1.68	-18.01	-24.85	-36.15	-2.86	-0.01	216.02
IPMTC3________	1.66	-18.14	-24.99	-36.30	-2.85	-0.15	215.89

Notes: Results of period 5 averaged over all agents, values represent the difference between the respective scenario and the baseline, total cultivated area in the model: 1100 ha

Also, the farmland cultivated with conventional leafy vegetables declines. It is clear that not only agents specialising in one land-use switch to IPM rotations of cabbage, lettuce, spinach and beans, but also that the increase in the area under IPM happens at the expense of all cash vegetable crops. However, agents growing chrysanthemums and/or roses seem to change their land-use only a little.

In contrast to the evaluation of the stand-alone tax interventions, which cause income losses greater than tax revenues to occur, and can be assessed with a cost-effectiveness analysis, the scenarios presented in this section involve income gains, for which a cost-effectiveness analysis is not required. The tax-effectiveness analysis is presented here, with the results very similar to those presented above. As can be seen from Table 44, it is also possible to simply evaluate the combined benefit of income change and tax revenues. The amount of pesticide reduction should however be the main criterion. Here, the proportional tax with and without compensation payment performs better than the flat taxes, as long as the focus is on toxic pesticide use, and this result confirms the previous evaluation outcomes for the taxation scenarios without IPM access. Figure 33 demonstrates that again compensation payments only have a pronounced effect for the high taxes scenarios, and that the relationship between these payments and the different tax levels is rather linear.

Considering that tax reimbursement does not allow reaching the IPM baseline income levels without tax intervention [IPM____________], the question needs to be asked if there is a more useful way of employing the tax revenues.

Table 44: Evaluation of policies for IPM + tax interventions

Scenario name	Δ income		Tax revenues	Policy costs	Total benefit	Δ pesticide use		Δ toxic pesticide use		Tax-effectiveness
						Av.	P. 5	Av.	P. 5	
	(%)	(1000 baht/hh)	(1000 baht/hh)	(1000 baht/hh)	(1000 baht/hh)	(%)	(%)	(%)	(%)	(Δ tox. pes. use/ tax revenues)
IPM	10.93	28.31	0.00	0.00	28.31	−5.53	−9.71	−7.81	−12.95	-
IPMTF1	9.61	24.99	1.35	0.00	26.33	−7.65	−12.08	−9.95	−15.05	7.38
IPMTP1	10.15	26.34	1.39	0.00	27.73	−7.02	−11.79	−10.16	−15.24	7.33
IPMTC1	10.24	26.58	1.39	0.00	27.96	−7.16	−11.47	−10.11	−15.04	7.30
IPMTF2	6.51	17.11	3.84	0.00	20.95	−10.79	−15.53	−12.68	−18.13	3.30
IPMTP2	6.97	18.23	3.91	0.00	22.14	−10.46	−15.80	−13.86	−19.47	3.54
IPMTC2	7.43	18.94	3.92	0.00	22.85	−10.46	−15.70	−13.82	−19.31	3.53
IPMTF3	4.34	11.55	6.09	0.00	17.64	−13.58	−19.34	−14.85	−22.02	2.44
IPMTP3	4.84	12.92	6.22	0.00	19.13	−12.90	−17.79	−16.32	−21.73	2.62
IPMTC3	6.04	16.00	6.22	0.00	22.22	−12.83	−17.86	−16.32	−21.81	2.62

Note: Averages over all agents and simulation periods, values represent the difference between the respective scenario and the baseline, for pesticide use reductions average (Av.) & period 5 (P.5) values are reported.

Therefore, in the following section several incentive mechanisms for promoting the diffusion of IPM will be assessed and which, when combined with a tax, can be compared to the reimbursement scheme.

Figure 33: Decision space for IPM + tax interventions

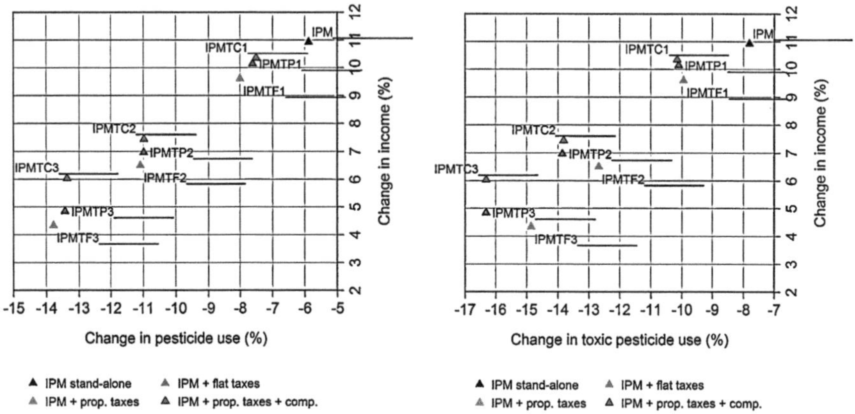

5.3.4 Impact of IPM adoption with adoption incentives

The following part of the scenario analysis deals with pesticide use reduction strategies that promote the adoption of IPM rather than penalise the use of pesticides. In this regard, three different incentive mechanisms are compared, each of which is, as before, simulated at three different levels of interventions (low, medium and high). Area payments [IPM___PA1______, IPM___PA2______, IPM___PA3______] provide temporary liquidity to farmers per hectare of land under IPM cultivation, in order to bridge the conversion period from conventional to integrated farming, during which lower yields might be expected. These payments are then gradually phased out over three years (for more details see Section 4.2). Price premiums [IPM______PR1___, IPM______PR2___, IPM______PR3___] are a second mechanism used to increase the attractiveness of IPM. Here, farm gate selling prices are increased by 1, 2 and 5% respectively. Finally, subsidies provided for IPM inputs are an important support measure that requires further study. The prices of biopesticides are then lowered by 20, 40 and eventually 60% to assess that effect on pesticide use, farm income levels and IPM diffusion. The IPM stand-alone scenario [IPM____________] is included in the figures and tables below as a reference.

Figure 34: Innovation diffusion for IPM + adoption incentives

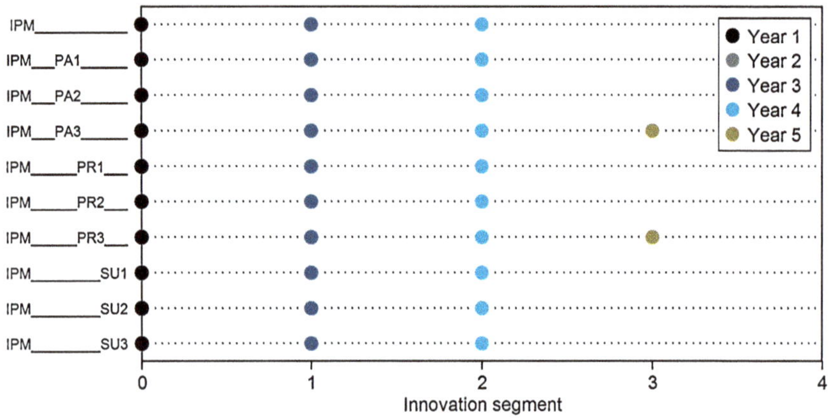

The diffusion of IPM, as shown in Figure 34, does not differ between the IPM reference scenario [IPM____________] and the low and medium area payment and price premium, as well as all the bio-pesticide subsidy scenarios. However, for the high support payment and price bonus level scenarios, access to IPM becomes available to segment 3 – the late majority – in year 5. The change in pesticide use is also highest for these two scenarios, reaching 10% for high price premiums but less than that for area payments.

Table 45: Simulated changes in pesticide use and income for IPM + adoption incentives

Scenarios	Description	Change in pesticide use			Change in income		
Name		(kg)	(kg/ha)	(%)	(1000 baht)	(1000 baht/ hh)	(%)
IPM____________	IPM access only	-882	-0.73	-5.53	42,216	28.31	10.93
IPM___PA1______	IPM + low area payment	-949	-0.78	-5.96	42,075	28.22	10.90
IPM___PA2______	IPM + med. area payment	-1,123	-0.92	-7.06	41,761	28.01	10.80
IPM___PA3______	IPM + high area payment	-1,335	-1.07	-8.07	25,700	17.24	6.62
IPM______PR1___	IPM + low price premium	-915	-0.75	-5.74	45,372	30.43	11.74
IPM______PR2___	IPM + med. price premium	-1,053	-0.87	-6.62	46,755	31.36	12.08
IPM______PR3___	IPM + high price premium	-1,612	-1.27	-10.09	66,088	44.32	17.01
IPM_________SU1	IPM + low biop. subsidy	-943	-0.78	-5.92	43,268	29.02	11.20
IPM_________SU2	IPM + med. biop. subsidy	-960	-0.79	-6.02	46,253	31.02	11.97
IPM_________SU3	IPM + high biop. subsidy	-1,032	-0.85	-6.49	46,758	31.36	12.07

Note: Averages over all agents and simulation periods, values represent the difference between the respective scenario and the baseline

As far as changes in incomes are concerned, farm agents are better off in all scenarios when compared to the baseline. They are also better off than in the IPM stand-alone scenario, with the exception of the three transitional area payment scenarios, and especially the high payment scenario. The bio-pesticide subsidy interventions are comparable to the other interventions at low levels, but create less pesticide reductions at the medium and high levels.

Figure 35: Distribution of pesticide use and incomes in the agent population for IPM + adoption incentives

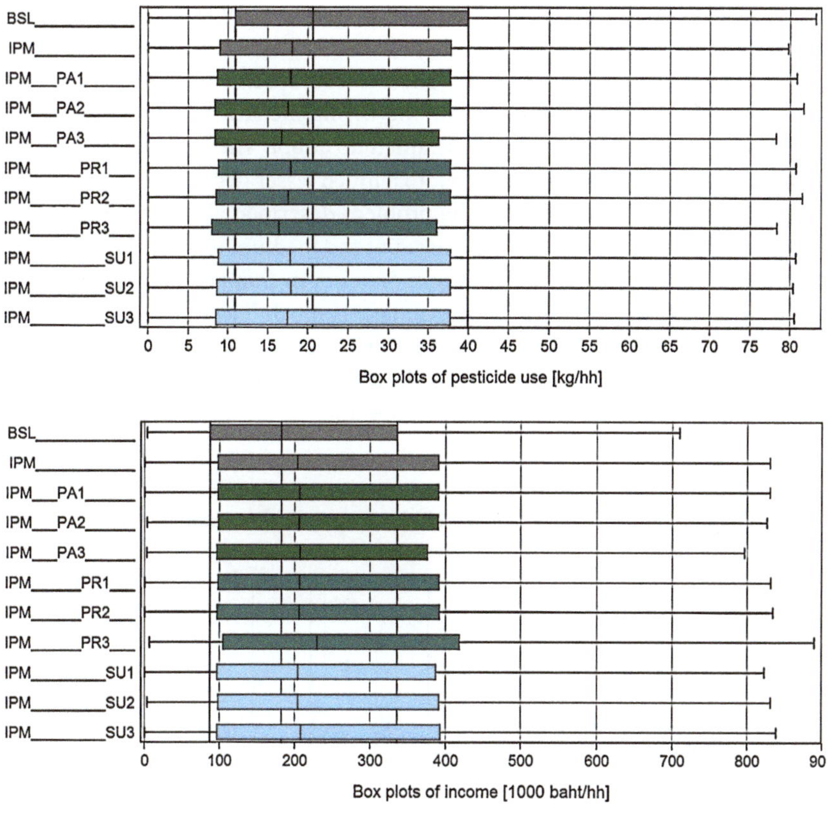

The box-plots in Figure 35 confirm the findings summarized in Table 45 and described above. The relatively poor performance of high area payments as opposed to the high price premium, which becomes evident when looking at the box plots, can be explained by the fact that agents do not have foresight about the duration of the payment. Therefore, they initially invest too much and later cannot maintain their plots under IPM, once the support falls gradually away or stops. The graph on the left in Figure 35 demonstrates that, when compared to the high price premium [IPM______PR3___], the high area payment [IPM___PA3______] results in many IPM adopters keeping their plots idle to a large or even full extent. As a result, the income increase is much lower since idle plots take land out of production that would otherwise be used, so generating no revenues. As IPM is implemented in the model as a perennial crop, in order to represent output and input requirements over time, as well as the required commitment by farmers, agents can only abandon IPM after six years. If facing insufficient funds, they cannot change to another land-use. It can thus be concluded that high area payments under the conditions created by the study MPMAS application create the wrong incentive, because too many agents that are financially very constrained do adopt. The high price premium also leads to some over-adoption, but to a much lesser degree. As shown in the graph on the right in Figure 36, the problem described above is no longer of great relevance once one takes into account the medium level of the price premium and area payment interventions. Here, even though adopters still keep some of their IPM plots idle, the percentage is much lower and the behaviour among adopters is more uniform across the scenarios. As opposed to the high intervention level, then according to these results, area payments at the medium level constitute a potential policy option.

Figure 36: Comparison of idle area under IPM between price premium and area payment scenarios

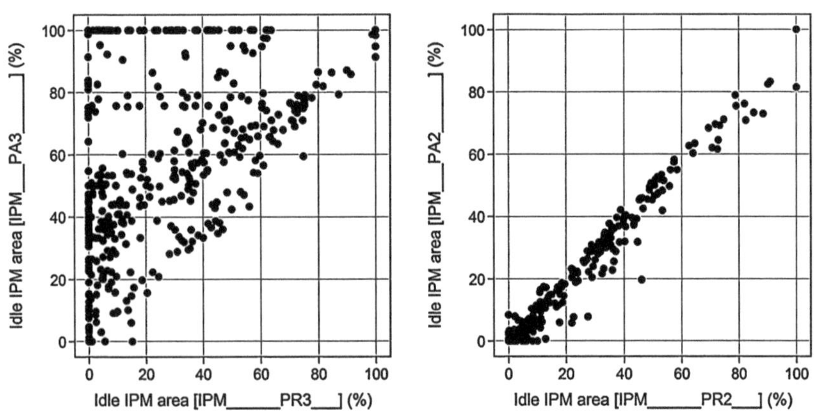

In terms of income distribution, the box plots above suggest that differences between the medium area payment [IPM___PA2______] and medium price premium [IPM______PR2___] are small. Figure 37 confirms this, but also shows that at the lower ranges of the distribution, income changes are slightly higher under the first scenario than under the second, while at the higher ranges the situation is vice versa.

Figure 37: Cumulative distribution of incomes for medium area payments and price premiums

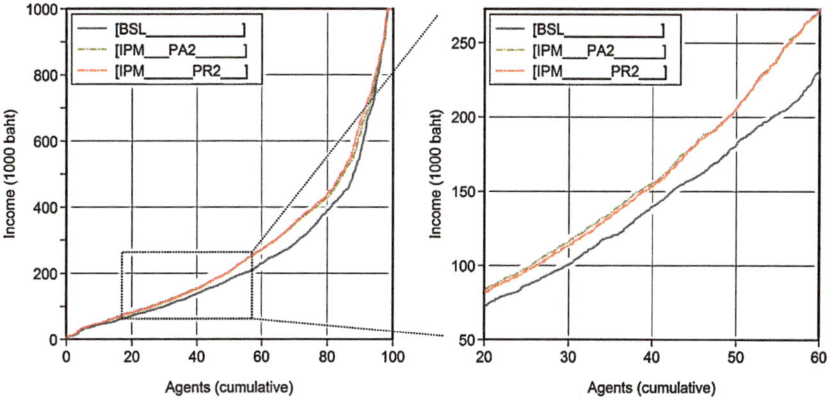

Land-use changes clearly differ across the scenarios in terms of the reduction of area under leafy and greenhouse vegetables. Table 46 illustrates that the area under IPM can be substantially increased, to ca. 190 ha, with high area payments or price premiums in place [IPM___PA3______, IPM______PR3___]. The other interventions, where the diffusion process does not reach the late majority, bring about a smaller change in the area under IPM, between the respective policy scenarios and the IPM reference scenario [IPM____________].

A final evaluation of the different IPM adoption promotion policies is required to first of all determine the most useful and practicable intervention if introduced as stand-alone policy, and also to analyse which interventions can be reasonably combined with taxes as well as with each other. Even though overall and toxic pesticide use are the most reduced by the high area payment [IPM___PA3______], as well as the high price premium [IPM______PR3___], in terms of expenditures, both scenarios create costs that are well beyond any potential tax revenues. In the case of the high proportional tax, as the scenarios in the previous section showed, the government could generate revenues of ca.

6000 Baht for each of the 1941 agent households (see Table 44). This implies that, apart from two scenarios [IPM___PA3______, IPM______PR3___], roughly all of the scenarios could be financed by tax returns. For reasons of not being realistic enough in terms of financing and practicability, and due to their undesired side-effects, these scenarios will be omitted in any further analysis.

Table 46: Land-use changes for IPM + adoption incentives

Scenario Name	Cereals (%)	Leafy veg. (%)	Greenh. veg. (%)	Other veg. (%)	Flowers (%)	Fruit tree (%)	IPM (ha)
IPM___________	−0.81	−19.31	−12.36	−34.71	−2.37	−1.17	207.25
IPM___PA1______	−1.14	−20.02	−12.77	−37.14	−2.37	−0.51	219.63
IPM___PA2______	−1.48	−20.36	−14.27	−38.88	−3.16	−0.08	230.61
IPM___PA3______	-9.74	−35.44	−27.03	−62.85	−6.64	−0.40	421.44
IPM______PR1___	−0.81	−19.65	−12.37	−36.77	−2.41	−0.52	218.60
IPM______PR2___	−1.31	−20.45	−13.21	−38.47	−2.82	−0.77	230.44
IPM______PR3___	-7.90	−34.32	−29.92	−72.21	−6.25	−0.20	413.92
IPM_________SU1	−0.73	−19.66	−12.58	−36.52	−4.06	−0.52	216.89
IPM_________SU2	−1.41	−20.23	−13.02	−37.51	−2.63	−0.69	224.04
IPM_________SU3	−1.37	−20.78	−13.55	−38.50	−2.80	−0.42	233.16

Note: Results of period 5 averaged over all agents, values represent the difference between the respective scenario and the baseline, total cultivated area in the model: 1100 ha

It might also be possible to have consumers bear the burden of the price premium; nevertheless, a price increase of 10% over the normal price for IPM vegetables still seems unviable. The farm gate selling prices for vegetables produced with IPM are already in the upper tercile of prices paid for the conventional leafy vegetables implemented in the model. A steep price increase might thus not be supported by the rather price-sensitive consumers in Thailand.

The cost-effectiveness values shown in Table 47 are a measure used to evaluate the policy costs of the IPM incentive, as against the amount of reduced, average toxic pesticide applications. The bio-pesticide subsidies turn out to be the most cost-effective interventions.

Table 47: Evaluation of policies for IPM + adoption incentives

Scenario name	Δ income		Tax revenues	Policy costs	Net benefit	Δ pesticide use		Δ toxic pesticide use		Cost-effectiveness
						Av.	P. 5	Av.	P. 5	
	(%)	(1000 baht/hh)	(1000 baht/hh)	(1000 baht/hh)	(1000 baht/hh)	(%)	(%)	(%)	(%)	(policy costs/av. Δ toxic pes.use)
IPM__________	10.93	28.31	0.00	0.00	28.31	-5.53	-9.71	-7.81	-12.95	-
IPM___PA1______	10.90	28.22	0.00	3.12	25.09	-5.96	-10.25	-8.55	-13.71	-0.37
IPM___PA2______	10.80	28.01	0.00	6.56	21.45	-7.06	-11.36	-9.76	-14.92	-0.67
IPM___PA3______	6.62	17.24	0.00	24.24	-7.00	-7.07	-19.22	-11.06	-25.24	-2.19
IPM_______PR1___	11.74	30.43	0.00	1.50	28.93	-5.74	-10.11	-8.26	-13.47	-0.18
IPM_______PR2___	12.08	31.36	0.00	3.85	27.50	-6.62	-11.04	-9.30	-14.69	-0.41
IPM_______PR3___	17.01	44.32	0.00	10.48	33.85	-10.09	-22.17	-13.10	-27.25	-0.80
IPM__________SU1	11.20	29.02	0.00	0.96	28.06	-5.92	-10.65	-8.28	-13.63	-0.12
IPM__________SU2	11.97	31.02	0.00	1.95	29.07	-6.02	-10.81	-8.60	-14.15	-0.23
IPM__________SU3	12.07	31.36	0.00	3.01	28.35	-6.49	-11.17	-9.29	-14.90	-0.32

Note: Averages over all agents and simulation periods, values represent the difference between the respective scenario and the baseline, for pesticide use reductions average (Av.) & period 5 (P.5) values are reported.

Apart from average numbers, Table 47 also informs about the pesticide use reductions during the last simulation period. Due to the progressive diffusion of IPM, the reduction in period 5 is considerably higher than the average reduction. An increasing pesticide use reduction can be assumed to perpetuate for ensuing periods, as long as new innovation segments gain access. However, once diffusion stagnates, the reduction of pesticide use doesn't change drastically. These simulations demonstrate that agents in the innovation segments with access to IPM adopt the innovation, but also continue to grow profitable conventional crops.

As far as the reduction in overall and highly toxic pesticide use is concerned, the medium area payment [IPM___PA3______] slightly outperforms the medium price premium [IPM_______PR3___] and the high bio-pesticide subsidy [IPM__________SU3], but income gains are higher for the latter two scenarios and the difference in pesticide use reduction is small. Considering the good cost-effectiveness of only 3,000 Baht per household for each percent of pesticide reduction, the [IPM__________SU3] scenario can be considered a good policy option. Figure 38, spanning the decision space, shows the respective performance of the incentive adoption scenarios. For instance, it becomes clear that the medium bio-pesticide subsidy [IPM_________SU2] achieves a similar pesticide use reduction as the low area payment [IPM___PA1______], but does so at a lower cost (see Table 47).

Figure 38: Decision space for IPM + adoption incentives, with scenarios [IPM___PA3______] & [IPM______PR3___] omitted

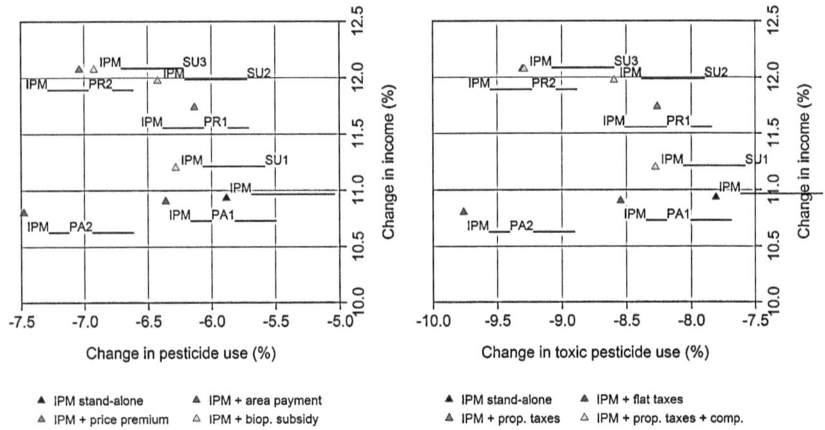

5.3.5 Impact of intervention mixes

In this section, several of the interventions that were analysed separately before, are now combined in order to produce a number of viable policy mixes. More combinations of policies were tested, but are omitted here. Most of the selected intervention mixes that are presented here constitute the most pragmatic options, with some scenarios being added for consistency and methodological considerations. This selection is listed in Table 48.

The colour code used in Table 48 as well as in the following tables and figures, classify the interventions in terms of the required expenditures. Green represents policy mixes that do not involve costs in excess of tax revenues. These scenarios can, in particular, be compared to the IPMTC3_________ set-up, which involves using the exact tax revenues to compensate farmers with lump sum payments. As regards the scenarios marked in yellow, the expenditures for measures in support of IPM slightly exceed the money received through taxes, but not by more than 10%, while the orange colour denotes substantial differences in both revenues and costs.

Table 48: Description of selected interventions mixes

Scenario name	Scenario description
IPM___________	IPM access only
IPMTC3________	IPM access + high proportional tax + compensation payment
IPMTP3PA1PR1___	IPM access + high proportional tax + low area payment + low price premium
IPMTP3PA1PR2___	IPM access + high proportional tax + low area payment + med. price premium
IPMTP3PA1___SU2	IPM access + high proportional tax + low area payment + med. bio-pesticide subsidy
IPMTP3PA1___SU3	IPM access + high proportional tax + low area payment + high bio-pesticide subsidy
IPMTP3PA1PR1SU1	IPM access + high proportional tax + low area payment + low price premium + low bio-pesticide subsidy
IPMTP3PA1PR1SU2	IPM access + high proportional tax + low area payment + low price premium + med. bio-pesticide subsidy
IPMTP3PA1PR2SU1	IPM access + high proportional tax + low area payment + med. price premium + low bio-pesticide subsidy
IPMTP3PA1PR2SU2	IPM access + high proportional tax + low area payment + med. price premium + med. bio-pesticide subsidy
IPMTP3PA2_____	IPM access + high proportional tax + med. area payment
IPMTP3___PR2___	IPM access + high proportional tax + med. price premium
IPMTP3___PR2SU1	IPM access + high proportional tax + med. price premium + low bio-pesticide subsidy
IPMTP3___PR2SU2	IPM access + high proportional tax + med. price premium + med. bio-pesticide subsidy
IPMTP3___PR2SU3	IPM access + high proportional tax + med. price premium + high bio-pesticide subsidy
IPMTP3______SU3	IPM access + high proportional tax + high bio-pesticide subsidy

Almost all of the high expenditure orange scenarios, apart from IPMTP3PA-1PR2___, allow for a higher diffusion of IPM than the remaining scenarios shown in Figure 39. For these remaining cases, as for most of the previously assessed scenarios, the diffusion process in year 5 reaches a stage where the late majority does not yet have access to the innovation. This means that diffusion is at a similar stage to the IPM reference scenario [IPM___________].

Figure 39: Innovation diffusion for IPM + intervention mixes

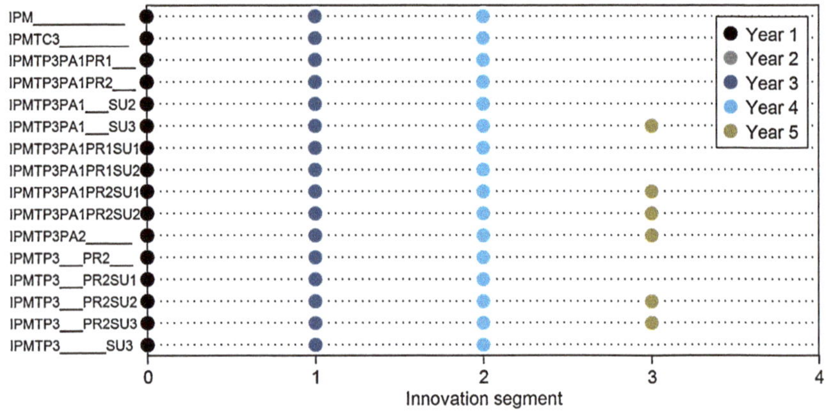

As far as Figure 40 is concerned, the pesticide use distribution is similarly affected in the IPMTC3_________ and the green and yellow policy mix scenarios. Apart from scenario IPMTP3PA1PR2___, in which IPM is not diffused to the late majority, the orange distributions are shifted somewhat further to the left, particularly at the 50th and 75th percentiles. Compared to the baseline, it becomes clear that altogether median pesticide use is more strongly impacted by the interventions than that at the 25th and 75th percentiles. In contrast, the impact on income is more severe at the 75th percentile. Green and yellow distributions are again similar to that of the IPMTC3_________ scenario, while the orange distributions generally allow the achievement of a similar income distribution to that for the IPM____________ scenario.

Land-use change depends heavily on the pattern of innovation diffusion present. If segment 3 of the innovation network gains access to IPM, as is the case in all but one of the orange scenarios, the area cultivated with IPM lies at above 400 ha. If not, the area under IPM does not rise beyond 250 ha, which can be seen by looking at the green and yellow scenarios in Table 49. The reduction in area planted with other crops follows a similar pattern for the two groups of scenarios. Interestingly, the area under greenhouse and other vegetables, such as onions and green beans, is much more affected than that under flower production. Further to this, the fruit tree area does not change much at all, which is related to the fact that as a perennial crop, the model does not allow the land-use to be replaced during its lifetime.

Figure 40: Distribution of pesticide use and incomes in the agent population for IPM + intervention mixes

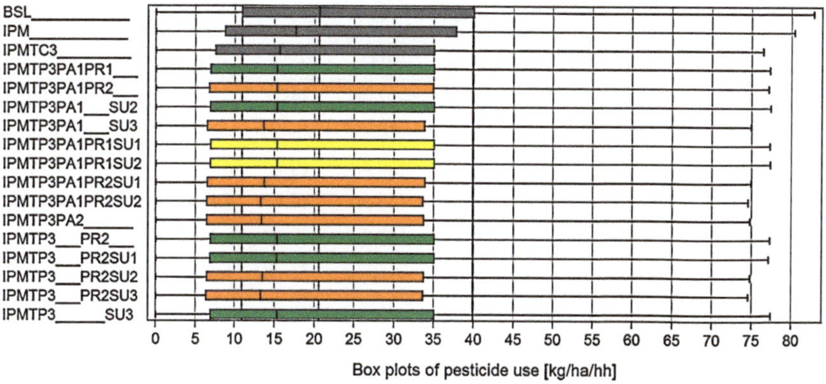

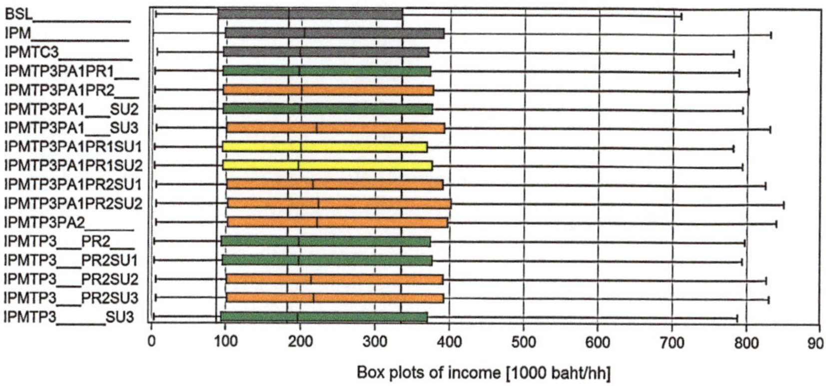

When evaluating the scenarios in terms of income changes and pesticide use reductions, it becomes evident that most of the green and most of the orange policy mix scenarios cluster together, as shown in Figure 41. Differences in pesticide use reduction among scenarios are very small for the green and yellow policy mixes, and only slightly more considerable for the orange intervention combinations. In fact, the yellow intervention mixes, as well as scenarios IPMTP3PA1PR1___ and IPMTP3PA1PR2___ perform comparatively less well as far as income changes are concerned. At the same time, the scenarios marked in yellow involve somewhat higher expenditures than the green scenarios, as Table 50 illustrates. Overall, when considering the more practicable green and yellow policy mixes, scenario IPMTP3___PR2SU1 performs best, achieving

Table 49: Land-use changes for IPM + intervention mixes

Scenario name	Cereals (ha)	Leafy veg. (ha)	Greenh. veg. (ha)	Other veg. (ha)	Flowers (ha)	Fruit tree (ha)	IPM (ha)
IPM_________	-0.81	-19.31	-12.36	-34.71	-2.37	-1.17	207.25
IPMTC3_________	1.66	-18.14	-24.99	-36.30	-2.85	-0.15	215.89
IPMTP3PA1PR1___	0.96	-19.11	-29.48	-40.23	-3.64	0.08	235.54
IPMTP3PA1PR2___	0.66	-20.24	-33.30	-40.67	-4.04	0.16	247.06
IPMTP3PA1___SU2	0.78	-19.68	-31.81	-40.43	-3.55	-0.08	241.62
IPMTP3PA1___SU3	-6.99	-33.49	-45.30	-73.20	-6.39	0.13	417.05
IPMTP3PA1PR1SU1	0.87	-19.75	-31.88	-40.50	-3.89	-0.24	242.61
IPMTP3PA1PR1SU2	-7.04	-33.42	-45.66	-73.16	-6.21	0.00	418.97
IPMTP3PA1PR2SU1	-8.04	-34.83	-47.43	-75.05	-6.78	0.16	441.29
IPMTP3PA1PR2SU2	-7.06	-33.62	-46.18	-73.58	-6.31	-0.04	427.90
IPMTP3PA2______	0.92	-19.18	-30.71	-40.37	-3.27	-0.39	236.06
IPMTP3___PR2___	0.91	-19.29	-29.26	-40.10	-3.76	0.29	237.46
IPMTP3___PR2SU1	0.67	-20.35	-32.14	-40.43	-4.83	0.11	244.32
IPMTP3___PR2SU2	-6.74	-33.38	-45.80	-73.93	-6.34	0.13	420.79
IPMTP3___PR2SU3	-7.40	-34.58	-47.40	-74.67	-6.86	0.13	433.62
IPMTP3______SU3	0.90	-19.47	-29.62	-40.01	-3.55	-0.01	239.93

Notes: Results of period 5 averaged over all agents, values represent the difference between the respective scenario and the baseline, total cultivated area in the model: 1100 ha

a reduction in highly toxic pesticide use of nearly 19%, while approximately matching the income level of the compensation scenario IPMTC3_________. Despite this, due to the uncertainty involved in the modelling process, scenarios IPMTP3PA1___SU2, IPMTP3___PR2___ and IPMTP3______SU3 also prove to be good alternatives, with expenditures being particularly low and so their cost-effectiveness being beneficial for scenario IPMTP3______SU3, which combines IPM and taxes with high bio-pesticide subsidies. In general, the cost-effectiveness ratio between policy costs and toxic pesticide use reduction is better for all yellow and green scenarios than for the compensation payment scenario IPMTC3_________.

As explained in the previous section for the adoption incentive scenarios, the reduction figures for the 5th simulation period exceed the average pesticide use reduction numbers, due to the gradual diffusion of IPM. Table 50 illustrates this for the mixed policy scenarios. In the case of scenario IPMTP3______SU3 for example, the average reduction in toxic pesticide use lies at 17.91%, while the reduction in year 5 already amounts to 24.43%. A reduction of nearly one quarter over a forecasting horizon of five years can be considered a significant policy impact.

Table 50: Evaluation of policies for IPM + intervention mixes

Scenario name	Δ income		Tax revenues	Policy costs	Net benefit	Δ pesticide use		Δ toxic pesticide use		Cost-effectiveness
						Av.	P. 5	Av.	P. 5	
	(%)	(1000 baht/ hh)	(1000 baht/ hh)	(1000 baht/ hh)	(1000 baht/ hh)	(%)	(%)	(%)	(%)	(policy costs/ av. Δ toxic pes.use)
IPM__________	10.93	28.31	0.00	−0.00	28.31	−5.53	−9.71	−7.81	−12.95	–
IPMTC3_______	6.04	16.00	6.22	−6.22	16.00	−12.83	−17.86	−16.32	−21.81	−0.38
IPMTP3PA1PR1___	4.65	12.37	6.08	−4.94	13.51	−14.42	−20.15	−18.07	−24.47	−0.27
IPMTP3PA1PR2___	5.69	15.01	6.03	−7.64	13.40	−14.88	−21.75	−18.58	−26.21	−0.41
IPMTP3PA1___SU2	5.51	14.54	6.06	−5.58	15.02	−14.73	−20.85	−18.17	−25.22	−0.31
IPMTP3PA1___SU3	9.34	24.70	5.84	−10.09	20.45	−16.96	−29.59	−20.89	−35.11	−0.48
IPMTP3PA1PR1SU1	5.23	13.82	6.06	−6.09	13.79	−14.61	−21.09	−18.20	−25.40	−0.33
IPMTP3PA1PR1SU2	9.06	23.96	5.83	−10.84	18.95	−17.15	−29.75	−21.00	−35.24	−0.52
IPMTP3PA1PR2SU1	10.12	26.73	5.83	−13.27	18.28	−17.10	−31.19	−21.46	−36.67	−0.62
IPMTP3PA1PR2SU2	11.09	29.26	5.79	−14.28	21.77	−17.45	−30.11	−21.00	−35.60	−0.68
IPMTP3PA2______	4.51	11.97	6.08	−6.71	11.34	−14.45	−20.38	−18.05	−24.74	−0.37
IPMTP3___PR2___	5.74	15.24	6.09	−4.11	17.23	−14.35	−20.09	−17.90	−24.40	−0.23
IPMTP3___PR2SU1	6.00	15.89	6.03	−5.29	16.64	−14.87	−21.65	−18.54	−26.15	−0.29
IPMTP3___PR2SU2	9.57	25.40	5.80	−8.14	23.06	−17.44	−29.91	−21.40	−35.31	−0.38
IPMTP3___PR2SU3	9.89	26.29	5.75	−9.71	22.33	−18.03	−31.02	−22.01	−36.52	−0.44
IPM__________SU3	5.52	14.68	6.09	−3.17	17.60	−14.38	−20.14	−17.91	−24.43	−0.18

Note: Averages over all agents and simulation periods; values represent the difference between the respective scenario and the baseline, for average pesticide use reductions (Av.) and period 5 (P.5), values are reported.

The orange policy mix scenarios, while are generally too expensive, might present a chance to further reduce pesticide use, under the condition that the price premium costs are passed on to the consumer. Scenario IPMTP3PA1PR2SU2 provides that the income level of IPM reference scenario [IPM____________] is reached; however, the costs are especially high and the prime criterion of pesticide use reduction is better met by scenario IPMTP3___PR2SU3. Combining a high proportional tax with IPM, a medium price premium and a high bio-pesticide subsidy is the most successful scenario in environmental terms. IPMTP3___PR2SU2 produces a similar reduction in pesticide applications as IPMTP3PA1PR2SU2, but with better cost-effectiveness, as Table 50 shows.

Figure 41: Decision spaces for IPM + intervention mixes

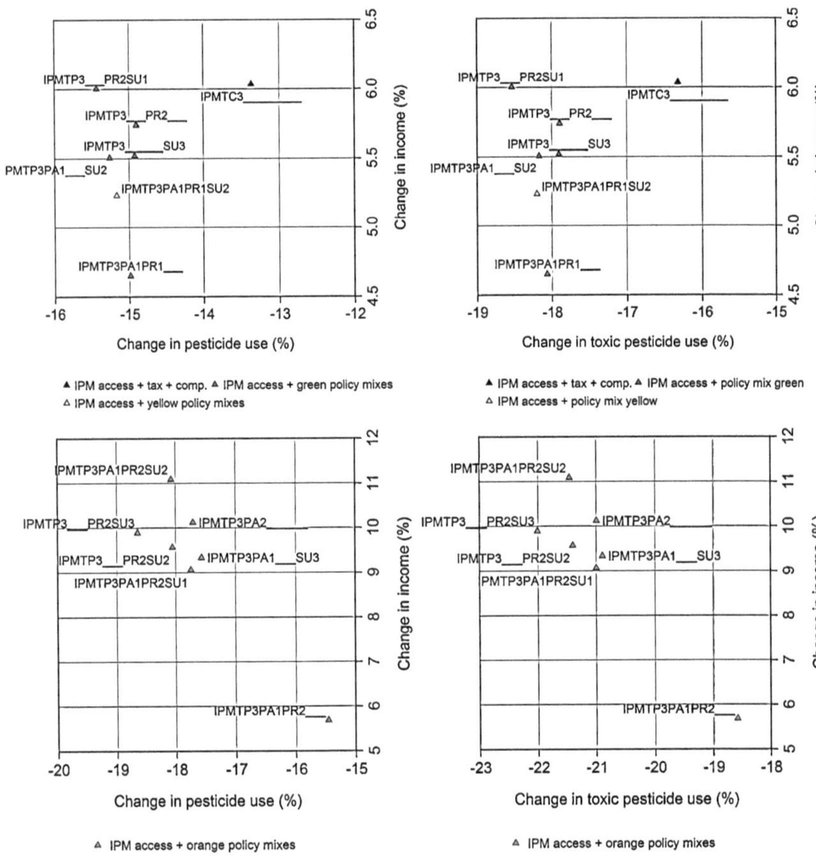

Considering the above scenarios, it can be deduced that overall differences between various policy mixes in the same cost ranges are not large. Nevertheless, it is possible to show, that spending the tax money on IPM promotion rather than redistributing it as a lump sum achieves higher pesticide use reduction rates. While mixes involving payments cannot generally be described as impracticable, price premium and bio-pesticide subsidy scenarios, and their combinations, tend to generate somewhat better results.

5.3.6 SWAT-based reductions scenarios for chlorothalonil and cypermethrin[7]

In contrast to the scenarios discussed previously, which are based on possible real-world policies, the scenarios presented in this section are based on inputs from the SWAT model, relying on a command-and-control mechanism rather than economic (dis-)incentives. Also, instead of examining the impacts on overall and toxic pesticide use, this assessment concentrates on two distinct pesticides. As such, the analysis differs considerably from the above policy scenarios, but can also complement them, as it shows what would happen if the policy-making focus was on two individual pesticides using a science-driven regulatory approach.

Figure 42 displays the effect of gradually reducing chlorothalonil and cypermethrin use on farm household incomes. At 50% of the baseline application of chlorothalonil, the average agent income is reduced by 11.8%. Decreasing the concentrations below the PNEC threshold (18% of the baseline use) reduces the income by 17.3%, from 249,149 Baht to 206,060 Baht. For cypermethrin, the more toxic substance, halving the application rate reduces farm incomes by 16.3%, while meeting the NOEC and PNEC thresholds reduces incomes by 36.5% and 38.3% respectively. The PNEC reduction, cutting down to 0.04% of the baseline dosage, is equivalent to an average income of only 153,849 Baht. For cypermethrin, the relationship between dosage reductions and income loss is linear. In contrast, for chlorothalonil, the more moderate dosage reductions affect average incomes less, while the higher reductions affect them more severely, as a linear correlation would suggest.

The box-plots in Figure 42 show that the incomes of the farm agents are not influenced to the same extent. The livelihoods of poorer farm agents are relatively less affected than those of the wealthier agents. For cypermethrin, at the 25th percentile the income loss is 5% for a 50% reduction and 31% when reduced to the PNEC level, whereas at the 75th percentile, incomes decrease by 19% and 42% respectively. Similarly, for chlorothalonil the dosage reductions have a less severe impact at the 25th percentile than at the 75th percentile. To reach PNEC concentrations, the respective income losses amount to 9% and 18%. However for chlorothalonil, a small proportion of agents run a deficit

7 Parts of this section have been submitted as: Bannwarth, M., Grovermann, C., Schreinemachers, P., Ingwersen, J., Lamers, M., Berger, T., Streck, T., subm. Non-hazardous pesticide concentrations in surface waters: Simulating application thresholds and resulting farm income effects. Journal of Environmental Management.

Figure 42: Impact on incomes and land-use for the baseline and different reduction scenarios

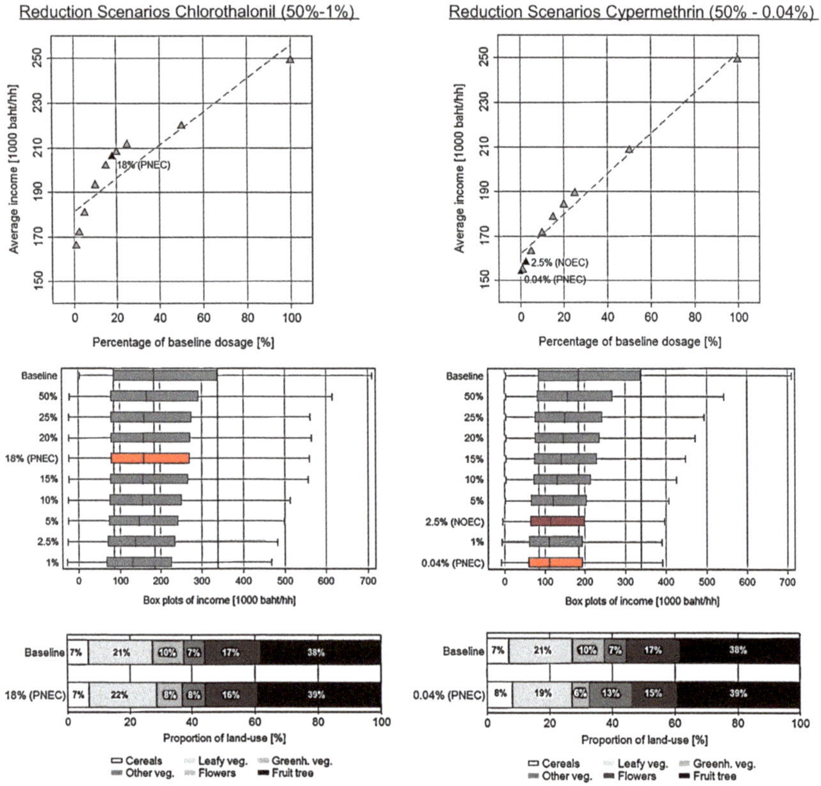

from the 50% reduction downward, whereas for cypermethrin an income deficit occurs only in the last three reduction scenarios. Since chlorothalonil is used a lot in the cultivation of roses, a perennial crop, agents can no longer continue growing roses under the respective model scenarios; they tend to abandon their fields because the model includes no option to replace perennials with other crops before its lifespan (seven years) is reached.

In terms of land-use changes induced by reduced pesticide use, the area used for growing greenhouse vegetables (bell peppers and tomatoes) is subject to the most marked decrease (after the area under flower cultivation), while the area under beans and onions (other vegetables) increases most noticeably. As can be expected, changes are more pronounced in the case of cypermethrin, since pesticide reduction under the PNEC scenario is more far-reaching. For instance the graphs at the bottom of Figure 41 show that the PNEC scenario for chlorothalonil entails a decline in the share of the area under greenhouse vegetables from 10% to 8%; for cypermethrin in the same scenario the share drops more substantially, to 6%. The share is not reduced to zero, because agents have the option to substitute pesticides and grow bell peppers and tomatoes, which require only a minimal or no application of either chlorothalonil or cypermethrin. Under the PNEC scenarios, land-use transformations as well as pesticide substitutions occurring in the different land-use categories involve changes in the mix of harvested products as well as lower yields, which then negatively impact upon income levels, as evidenced above.

5.4 Key lessons learned for policy-making

The issues of food safety and pesticide risk reduction continue to be high on the agenda of Thai policy-makers, which raises the question as to what should be recommended in order to maximise the potential realisation of pesticide use reduction policies. This thesis analyses a range of economic instruments that can help to tackle the problem of heavy pesticide use, though not all of these instruments are equally effective, practicable or relevant. The following section clarifies which policies are best suited and explains why.

First of all, it should be noted that both a 50% flat tax as well as a tax that increases by toxicity at 40%, 50%, 60% and 70% increments result in a moderate 7% to 8% reduction in pesticide applications, which can be seen in Table 51. When weighing up one tax scheme against the other, it is important to point out that a flat tax could have an adverse environmental effect. The reason for this is that such a tax could lead to an increase in the use of cheap broad-spectrum pesticides, which are generally more hazardous than the more expensive but less toxic narrow-spectrum pesticides. In the taxation simulation experiments, a rise in toxic pesticide use for flat taxes when compared to the baseline is not observed, but it becomes clear that the use of less toxic pesticides (WHO III & WHO U) diminishes to a greater extent than the use of toxic pesticides (WHO I & WHO II). As expected, the effect of the proportional tax differs when compared to the flat tax, but the generally low reduction of WHO I pesticide use is notable. Due to their association with highly profitable cropping activities in the model, WHO I applications are reduced less than those of the WHO II pesticides. Surprisingly, the use of not listed substances (WHO NL) increases under both the proportional and the flat tax scenarios, suggesting that they fulfil an extremely important purpose for farmers, and also that they substitute for other pesticides once taxes are introduced.

Table 51: Flat vs. prop. tax: Reduction of pesticide use by WHO toxicity class

Scenario		WHO I (%)	WHO II (%)	WHO III (%)	WHO U (%)	WHO NL (%)	All pesticides (%)	Tax revenue (1000 baht/hh)
___TF3_______	High flat tax	−3.17	−8.95	−9.65	−10.79	2.96	8.47	6.56
___TP3_______	High prop. Tax	−4.45	−10.49	−8.56	−8.49	3.30	7.34	6.74

Note: Averages over all agents and simulation periods; values represent the difference between the respective scenarios and the baseline

Lesson learned 1: High taxes achieve a moderate reduction in pesticide use, not exceeding 7% to 8% over the simulation period. Tax impacts do not get progressively greater with time. While flat taxes, as expected, show a lack of specificity, proportional taxes target toxic pesticide use somewhat better. The difference between both types of taxes is not large.

Tax revenues can be employed to either directly compensate farmers through lump sum payments or promote less pesticide-intensive production practices. Contrary to the tax-compensation scheme, the latter option has a clear temporal dimension, since impacts can be assumed to become more significant over time. As far as the policy measures in support of integrated pest management are concerned, discussions about the results of this simulation with experts on pesticide policy from Kasetsart University in Thailand, provided information as to what interventions can be considered most suitable for actual implementation. According to the experts, a subsidy for bio-pesticides would most likely be put into practice, since it fits the existing policy framework which already allows for the subsidising of various agricultural inputs. Expert opinion favours subsidies over a price premium, since the latter would require a new policy infrastructure to be in place, while the financial burden is unlikely to be accepted if shifted to consumers. Price-sensitive consumers in Thailand would be reluctant to pay a price premium for vegetables produced using with IPM practices.

The cost-effectiveness analysis carried out for different simulation results (see Section 5.3.5) also comes out in favour of advocating a bio-pesticide subsidy over other policy measures. The same reduction of pesticide use is achieved using a 5% price premium involving average costs of 4,110 Baht per household, and a 60% subsidy for bio-pesticides for which average costs only lie at 3,170 Baht per household. This cost is well below tax revenues, which leaves room for the augmentation of bio-pesticide use and the related, additional government expenditures. However it should also be noted that transaction costs are not considered within the present analysis. The third economic instrument analysed by this study, transitional area payments, are by design different to the other two interventions, since their duration is limited. The advantage of a temporary policy intervention is its ability to boost the adoption of an innovation initially without creating costs in the long run. For a profitable innovation such as leafy vegetables under IPM which involves some lower yields during the conversion period, the level of adoption initially achieved is likely to be sustained beyond the point when the payment ceases. Therefore, transitional area payments can be a valuable instrument – especially at the project level – to use in order to speed-up innovation diffusion while avoiding long-term costs. Over the five year simulation period within the model here, it is shown that such payments are comparatively less cost-effective than the other two incentive instruments, and that they can create a misleading motivation to adopt. Particularly the high levels of the payment induce farmers to adopt who don't have the means to carry on with IPM production once the payment has been phased out. This effect might be observable with any transitional policy measure implemented within the model to promote IPM, because agents cannot abandon IPM in order to grow other crops in the short run.

The left radar plot in Figure 43 shows the relative performance of the compensation payments scenario [IPMTC3__________], the 5% price premium scenario [IPMTP3___PR2___] and the 60% bio-pesticide subsidy scenario. All of these scenarios combine IPM access, a high proportional tax and a redistribution of tax revenues, and belong to the green policy mixes, which are capable of being funded by tax revenues. Apart from the positive income increase, the bio-pesticide subsidy out-performs the other two scenarios in all respects. On the right hand side of Figure 43, the radar plot shows three of the orange policy mixes with similar policy costs. However these costs exceed tax revenues. It becomes clear that the scenario that combines a 5% price premium with a 60% bio-pesticide subsidy [IPMTP3___PR2SU3] performs best; it achieves the highest reductions in overall and toxic pesticide use among all the simulated scenarios. Nevertheless, when compared to the scenarios displayed in the left hand radar plot, the cost-effectiveness is lower. As mentioned before, as well as being expensive and less cost-effective, more complicated policy mixes are also not easily implemented.

Lesson learned 2: The introduction of IPM in combination with bio-pesticide subsidies that are financed by a pesticide tax, turns out to be the most cost-effective and practicable pesticide use reduction strategy. With a 60% subsidy, pesticide use is reduced by ¼ by the end of the model simulation.

Figure 43: Radar plots comparing green (left) and orange (right) policy mixes (Variables are normalised, 0–1)

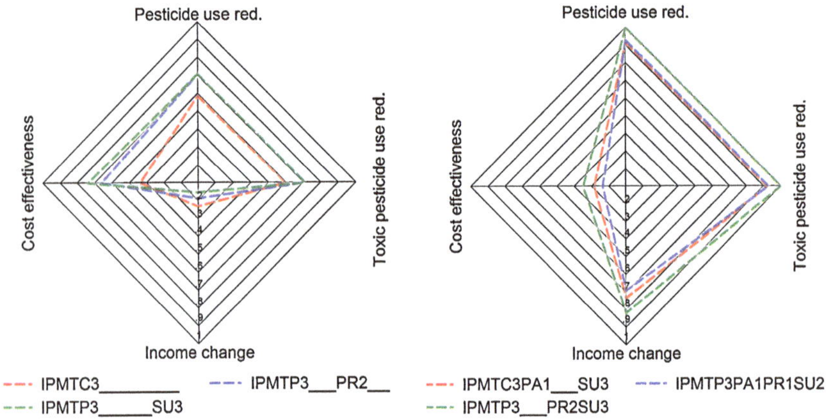

Across the spectrum of high and low polluting agents, the reduction in pesticide use, including highly toxic pesticide use, is not entirely uniform. Figure 44 illustrates the effect of dividing the agent population into 10 equal polluter groups, i.e. polluter deciles ranging from low polluting agents who apply few pesticides, to highly polluting agents who spray heavily. For scenario ___TC3_________, which involves a high proportional tax and compensation payments, pesticide use is least reduced in the lowest two deciles and most substantially reduced in the highest two. However, the relationship between the amount of baseline pollution and pesticide use reduction is not completely linear. It can be said that the heavy polluters rely greatly on toxic pesticide inputs and so are hardest hit by the tax.

For the other scenario represented in Figure 44, IPMTP3_______SU3, the percentage reduction in toxic pesticides use is lowest for the first decile and highest for the ninth. In the other deciles, the level of reduction is very similar. The dynamics of pesticide use reduction in the agent population is determined by the adoption of IPM, which in turn is driven by the innovativeness of individual agents. Average innovativeness is found to be highest in the ninth percentile. The fact that reduction in the last decile, representing the top polluters, is lower than might be expected, can, apart from innovativeness, be explained by the input-intensive, profitable crop mix that these wealthy and heavily polluting farm agents grow. They are, as a result, less inclined to swap over to the growing of leafy vegetables with IPM practices.

Figure 44: Pesticide use reduction and income changes across high and low polluting agents in period 5

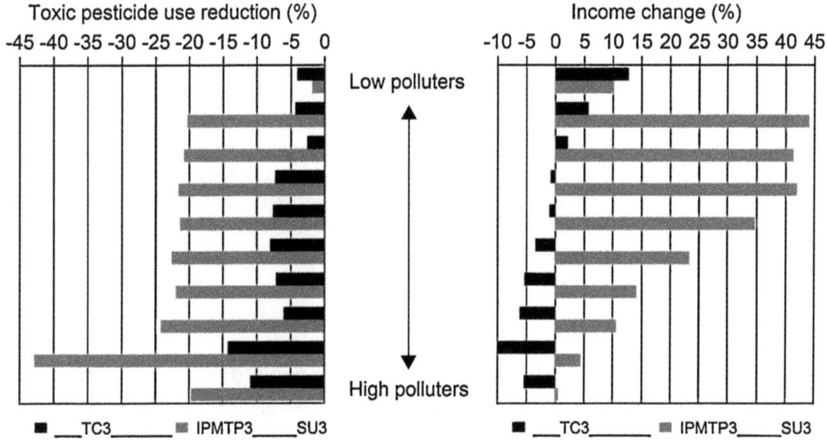

As shown in the results chapter, the degree of pollution is closely related to the income situation of farm agents, with higher pesticide pollution levels largely caused by the more wealthy agents. The high polluters are those that are burdened more significantly by a tax and benefit less from the introduction of IPM in combination with the subsidy. The negative income effect of the tax, which cannot be fully abated by the compensation payment, disappears under the scenario IPMTP3______SU3, as none of the polluter groups incur income losses on average. The highest income gains occur in the deciles in which agents switch from relatively lower-value production to the high-value IPM cropping activities, demonstrating that poorer segments of the population can benefit from the introduction of IPM.

For the IPMTP3______SU3 scenario, Figure 45 further reveals the land-use dynamics behind the trends seen in Figure 44. The low polluters are those farm agents growing a lot of rice and maize (Cereals), and also chrysanthemums (under Flowers). As pollution levels increase, leafy and greenhouse vegetables constitute more important land-uses, but the area under IPM increases, since conventional vegetable production is partly replaced by IPM production. This is especially the case for the 9th decile, whereas in the 10th decile, for the most severely polluting agents, IPM plays a minor role since growing the conventional activities (mainly bell peppers, tomatoes, cabbages and roses) produce generally higher returns for these agents.

Figure 45: Land use in each of the 10 polluter groups, for scenario [IPMTP3______SU3] in period 5

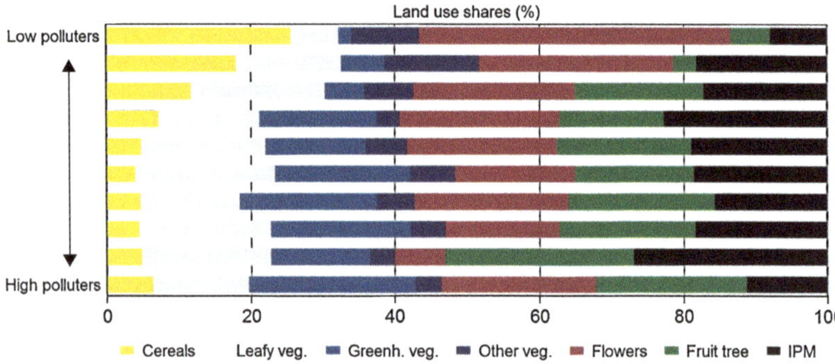

Lesson learned 3: Factors determining pesticide use reduction are distinct in the tax scenarios (monetary disincentives) and IPM adoption scenarios (innovativeness, relative profitability of existing crops). Income changes among farm agents are a function of the respective financial viability of the conversion to IPM practices.

Once the IPM adoption potential in a particular innovation segment has been reached, only access to IPM by the next segment can produce a substantial rise in pesticide use reduction amounts. If additional financial resources were available, it could be argued that rather than spending these on a mix of adoption incentives, it would be of more benefit to increase the bio-pesticide subsidy in order to boost the diffusion of IPM in a cost-effective way. Therefore, it would be interesting to test if higher subsidies can trigger a jump from the early to the late majority as far as IPM access is concerned. As Table 52 shows, a subsidy higher than 60% is able to boost the diffusion of IPM, and so lead to a large increase in pesticide use reduction. However it can also be seen that the difference between the 80% and 100% subsidy is modest, since the 80% subsidy is already sufficient to motivate most potential adopters in segment 3 - the late majority adopter group – to select IPM, and so substantially decrease their pesticide applications. When compared to some of the intervention mixes described in Table 50, which perform similarly in terms of pesticide use reduction, such as scenario IPMTP3PA1PR1SU2 for example, the cost-effectiveness as well as the practicability in terms of realisation of an 80% bio-pesticide subsidy is superior. Therefore, if aiming at higher pesticide use reduction figures and with sufficient funding available, it is more reasonable to step-up the bio-pesticide subsidy as an IPM adoption incentive, rather than adding other incentive mechanisms. In this regard, it should be noted that the policy costs reported in Table 52 are averages over the five simulation periods.

Table 52: Impact of increase in bio-pesticide subsidies

Name	Scenarios	Δ income	Tax revenues	Policy costs	Δ pesticide use		Δ toxic pesticide use		Innov. access	IPM area
					Av.	P. 5	Av.	P. 5	P. 5	P. 5
	IPM + tax +	(%)	(1000 baht/ hh)	(1000 baht/ hh)	(%)	(%)	(%)	(%)	(Adopter group)	(ha)
IPMTP3_______SU3	Bio-pesticide subsidy 60%	5.52	6.09	3.17	–14.92	–20.14	–17.91	–24.43	Early majority	239.93
IPMTP3_______SU4	Bio-pesticide subsidy 80%	8.70	5.86	5.38	–17.42	–28.98	–20.72	–34.33	Late majority	414.53
IPMTP3_______SU5	Bio-pesticide subsidy 100%	8.72	5.78	6.89	–18.29	–30.41	–21.51	–35.53	Late majority	425.65

Note: Averages over all agents and simulation periods, values represent the difference between the respective scenario and the baseline, for pesticide use reductions average (Av.) & period 5 (P.5) values are reported.

While an 80% subsidy can be financed through tax revenues during the simulation run, funding sources in addition to the tax would be required in the long run, since high IPM adoption rates entail rising policy costs.

Lesson learned 4: Extra funds are best spent on raising bio-pesticide subsidies to boost further IPM diffusion. An 80% rather than 60% subsidy produces an added 10% of pesticide use reduction. However, as demand for bio-pesticides increases with IPM diffusion, costs rise over time.

The approach of using scientifically derived reduction targets for particularly important pesticides contrasts with the general policy-based reduction scenarios for all pesticides. This approach is motivated by the use of eco-toxicological information as the basis of regulatory intervention, rather than by simulating and analysing the impact of potentially practical policies. In this context, the NOEC and PNEC thresholds for cypermethrin application require drastic reductions of this important active ingredient. The impact on the income of farm agents is huge; therefore, the related scenarios are less realistic than a pesticide tax. Also, the monitoring costs raise doubts about the viability of implementing strict eco-toxicological guidelines for individual pesticides such as chlorothalonil and cypermethrin.

6. Discussion and conclusion

6.1 Strength and weaknesses of the econometric analysis

A novelty of the econometric analysis shown in this paper is the inclusion of pesticide externalities when quantifying pesticide overuse levels. The PEA tool is straightforward to apply if farm-level data on active pesticide ingredients are available. Currently it appears to be the only available tool able to do this. However, several weaknesses in the methodology need to be considered, as discussed by Praneetvatakul et al. (2013), meaning there is room for improvements to this methodology.

The production function approach is based on standard micro-economic theory and assumes that farm decision-making is guided by a profit-maximizing motive. Nevertheless, in reality, there are other motivations for farm decision-making, but including these would make the calculation of economic optima very complex, and would require an unrealistically high amount of farm-level data. Therefore the idea of profit maximization is a necessary simplifying assumption for ease of computability.

The model found that 78% to 79% of the total quantity of pesticides applied can be labelled as overuse from a private point of view. This implies that farmers, even without considering externalities, are spraying excessively and inefficiently, and could increase their profits by applying fewer pesticides. The marginal value product was estimated to be close to zero for the majority of observations, which confirms the results of Praneetvatakul et al. (2003), who used the same functional form and also found the marginal value product of pesticide use in rice farming to be close to zero. Studying pesticide overuse by vegetable farmers in Nepal, Jah and Regmi (2009) likewise used a Cobb-Douglas function with an exponential abatement specification, and found that 70% of pesticide use was above the private optimum, a finding similar to that of this study.

The analysis found only a one percentage point difference between private and social levels of overuse for this case study of upland horticulture, which appears surprising at first glance. The difference is small because the optimum level of control is reached at a relatively low level of pesticide use, that is, the function is steep, as illustrated in Figure 3. Most farmers produce at the flat-end of the production function, where the marginal value product approaches zero. It is not possible to conclude from this result that including pesticide externalities is not important, but the results do suggest that for situations where pesticide overuse is dramatic, internalizing pesticide externalities into the retail price of pesticides, for instance through an environmental tax on pesticides, might not be an effective policy instrument. In other words, a marginal value product of pesticides that approaches zero suggests that farmers are not much influenced by the costs and returns of pesticides when deciding whether and how to use them; if they were, they would bring the marginal value product closer to unity. Though the findings here are from one location only and at this stage of the research the findings cannot be generalized for the whole of Thailand, it does confirm previous studies, which showed the demand for pesticides to be very inelastic (Falconer, 2000; Pina and Forcada, 2004; Skevas et al., 2012).

A possible explanation for the high rate of pesticide use is that farmers in the study area have few alternatives to synthetic pesticides due to a lack of knowledge about available methods to manage pests in an integrated manner. Government policies, such as tax exemptions and subsidized credit for chemical inputs, keep the price of pesticides low and contribute to application rates above optimal levels. Farmers are also 'locked in' to a system of unsustainable pest control, because of the real or perceived economic losses of switching their growing practices (Wilson and Tisdell, 2001). Non-synthetic methods of pest control are only applied on 8% of the planted area, with 77% of the farms solely depending on synthetic pesticides. The development and dissemination of integrated pest management is in its infancy in Thailand, and investment in these might have a more substantial and long-term effect on reducing pesticide use than simply removing pesticide subsidies or introducing a standalone tax. To test the validity of this statement, in this study it was first necessary to examine what determines the knowledge and adoption of agricultural innovations in the context of upland farming in northern Thailand. For this purpose the recently introduced public GAP standard offered a suitable research object. Discrete choice models are commonly found in the literature to model adoption decisions (Neupane et al., 2002; Thapa and Rattanasuteerakul, 2011; Wollni and Zeller, 2007). However, biases from omitted variables and poor model specification can produce inconsistent results and flawed conclusions (Lindner, 1987), so the econometric model used in this research was carefully selected to fit the nature of the innovation. Since knowledge of GAP certification is limited to only a proportion of farm households in the survey, a probit model with sample selection was used. Also, a range of network characteristics were included in the regression. The comprehensive set of explanatory variables, and the econometric model itself, were instrumental in avoiding biased results.

Generally, the regression outcomes suggest that efforts to promote learning from farmers who already know the innovation and/or are members of extension and producer organisations are key to increasing participation in the GAP programme. Education of farmers is also a crucial factor in this regard. The findings confirm the role of exposure to individuals with knowledge of the innovation, which is a cornerstone of innovation diffusion theory (Valente, 2005) and is also considered important within research on the adoption of agricultural innovations (Abadi Ghadim and Pannell, 1999). While not explicitly quantifying exposure, a study on the adoption of organic agriculture in Thailand identified motivation by community members and by farmer groups as crucial determinants of adoption decisions (Thapa and Rattanasuteerakul, 2011). The results here show that besides links to informed farmers, links to outside actors are highly significant in explaining innovation adoption; the main outside actor being the Royal Project, a Thai extension and producer organisation. Studying the adoption of sustainable resource management in Ethiopia, Wossen et al. (2013) affirmed that external sources of information for extension provision play a crucial role in enhancing adoption. Wollni and Zeller (2007) found that participation in cooperatives has a positive impact on innovation adoption in their case study of specialty coffee bean cultivation. They also pointed out that this provides evidence of successful advisory services being provided by cooperatives, but that this finding needs to be interpreted with caution, for the organisations might promote the innovation only among those farmers they consider more likely to adopt it. Another point to consider is that it is not the lack of adoption but a lack of tangible sustainability improvements due to the public GAP standard in Thailand that is a major concern. Low motivation rates among farmers and little understanding of control points have prevented the GAP standard from delivering on its plan to reduce quantities of applied synthetic pesticides (Schreinemachers et al., 2012).

In the case of Thailand it can be said that, as well as advice, farmers participating in the Royal Project are given a clear incentive to reduce pesticide use, because the products it certifies can be sold through special marketing channels. Also, farmers growing GAP certified vegetables receive slightly higher prices in general (Kasem and Thapa, 2012) than those who do not. The economic dimension of adoption is reflected in the innovation implementation phase, which according to Rogers (2003) follows the decision to accept or not accept the innovation. Lindner (1987) provided compelling empirical support for the emerging consensus that the final adoption choice is consistent with a producer's self-interest. In this sense, the econometric model predicts adoption probabilities that provide insights into what determines adoption, while the final decision on the extent of adoption for profit-maximising farmers depends on the profitability of the innovation. In this regard, the econometric analysis of adoption can be gainfully complemented by agent-based modelling with MPMAS, which uses the analysis outcomes to simulate innovation diffusion to agents, and in a second step then simulates constrained profit-maximisation by agents.

6.2 Strength and weaknesses of the MPMAS application

Overall, the temporal and spatial dimensions of the model, the combination of social network and optimisation dynamics, as well as the analytical gains to be made by being able to simulate impacts on a heterogeneous population of farm agents, are original features of MPMAS. The strengths of the present MPMAS application are that it harnesses these features and applies them to a new field of study. Agent-based modelling has so far not been used to assess crop protection policies. It increases the complexity of modelled processes and helps avoid problems of over-specialisation and aggregation bias inherent in previous research using representative farm MP modelling (Falconer, 2000) or aggregate sector MP modelling (Jacquet et al., 2011). The results obtained with MPMAS illustrate the adjustments and reactions of individual farm agents to crop protection innovation and pesticide policy interventions, permitting an analysis of impacts for different polluter groups, including a more detailed representation of the dissemination of IPM among agents. The incorporation of network constraints alongside optimisation in a multi-agent system generally distinguishes MPMAS from other bio-economic farm models used to assess innovations and responses to policies (Janssen and van Ittersum, 2007). Also, compared to a rule-based multi-agent system such as the CATCHSCAPE model developed for northern Thailand (Becu et al., 2003), MPMAS stands out, as optimising agents with innovation access can evaluate adoption more effectively against the full range of existing cropping activities.

The level of agent access to an innovation is defined by the stage of innovation diffusion and by the agent's innovativeness. This presumes that the discrete innovativeness variable, which is calculated for each agent, can capture a range of factors such as social position, farm characteristics, and risk perceptions and attitudes. Therefore, the default random allocation of innovativeness in MPMAS has recently been replaced by a more refined direct assignment approach, which goes beyond previous improvements to the default set-up in the MPMAS application for Thailand or Vietnam (Quang et al., 2014; Schreinemachers et al., 2009). The validation used here clearly demonstrated the added value of the current approach, since the more sophisticated representation of innovativeness evidently delivers more accurate results. It has been argued that risk perception is often absent from adoption research (Abadi Ghadim and Pannell, 1999), yet the method of calculating agent innovativeness used in this study takes into account risk by applying the criterion of the time lag between introduction and individual adoption. This can be understood as an indicator of risk aversion and complements the estimated adoption probabilities when predicting innovativeness. Another concern raised by Abadi Ghadim and Pannell (1999) refers to the learning process, which according to them consists of a trial and error period in which farmers evaluate an innovation. The lack of learning-by-doing limits the scope of action of farm agents and constitutes an important shortcoming of the innovation diffusion approach, as modelled in this MPMAS application. It is the result of implementing IPM for leafy vegetables as a perennial crop which prevents agents from abandoning it in the years after adoption, though they can keep the land under IPM idle. The disadvantage of this needs to be weighed against the advantage of using the perennial crop feature for IPM, which allows the model to represent changing yields over the lifetime of an innovation. It can also be put forward in favour of the current implementation that once an investment in IPM infrastructure is carried out, a change to IPM practices cannot be easily reversed within one year.

The spatial set-up of the Mae Sa watershed MPMAS application takes into account landscape components, in particular crop suitability in relation to slope, but is based on a rough approximation of the location of farms in space due to the stratified random allocation of farmsteads and plots. Moreover, distance is a spatial factor that might determine innovativeness or market access, but is neglected in the study model. Due to the compact size and relative homogenous commercialisation of the watershed, this can be regarded as an acceptable simplification of the spatial aspect of agriculture within the study application. As the watershed varies greatly in terms of topography, growing activities in the model are specified according to slope type. For these activities, the relationship between pest pressure, pesticide use and avoided damage to crop yields is however only very generally represented. The production functions with exponential damage control specification used to identify empirical observations for the model's growing activities are a means of characterising the indirect effect of pesticides on productivity, i.e. the abatement of pest damage. However, theory offers little guidance on how exactly damage functions should be specified, with the choice of specification having a considerable impact on pesticide productivity estimates (Carrasco-Tauber and Moffitt, 1992; Sexton et al., 2007). Also, the damage control approach doesn't capture many underlying biological processes that determine pest control and crop damage, such as resistance build-up and predator-prey relationships. Shankar and Thirtle (Shankar and Thirtle, 2005) pointed out that econometric analyses of pesticide productivity typically fail to incorporate entomological information and detailed stage-by-stage data on pest infestation and pesticide applications. Information on the extent and timing of pest infestations and pesticide applications is rarely available though, as was the case for this research, which instead used cross-sectional data. On top of this, to estimate pesticide productivity, many different active ingredients with various effects and toxicity levels have to be grouped together. Nevertheless, given the limitations imposed by recall farm survey data, the damage control framework proposed by Lichtenberg and Zilberman (1986) provides a more accurate analysis of pesticide productivity when compared to traditional production function analysis (Jah and Regmi, 2009; Shankar and Thirtle, 2005). Moreover, to the author's knowledge, there are no alternative and well-established methodologies available to estimate pesticide productivity.

In their assessment of economic incentives for pesticide use reduction, Skevas et al. (2012) extended the standard econometric model of production function with exponential damage control term, in order to capture the effect of environmental spill-overs such as biodiversity loss. They used data from three cropping seasons and focused on potato farmers only. Predicted coefficients served to model the impact of taxes or quotas on reducing pesticide use, which was split into two toxicity classes. Unlike MP-based models, such a statistical modelling approaches cannot fully represent the substitutions made between a big range of active ingredients. Falconer and Hodge (Falconer, 2000) confirmed that an important reason, why actual responses may be higher than theoretically predicted, is related to the assumptions and reductionist approaches used when modelling, especially with regard to the range of options available to producers. Falconer and Hodge's seminal case study of pesticide use reduction policies in the UK is based on a representative MP farm model and focuses on taxes and levies. The data used for their model stems from experimental trials for 12 crops, those which serve as approximations of conventional and low-input farming production activities (Falconer, 2000, 2001). Similarly, an MP-based pesticide policy model for the French agricultural sector developed by Jacquet et al. (Jacquet et al., 2011) is built on agronomic trial results and expert knowledge. In contrast to these two models, the MPMAS model developed here exhibits a more data driven model set-up, because it contains actual empirical observations with varying pesticide observations. The complexity of the horticultural farming system is represented by 513 growing activities, which are related to 85 active ingredients in total. Instead of empirical observations, it would have been possible to resort to expert opinion in order to parameterise the cropping activities for the model. While the entomological and ecological aspects of crop-pesticide interaction could have been represented more accurately, the empirically-grounded model implementation more closely reflects the actions of the farmers themselves, who are the decision-makers and the party affected by policy interventions. This allows the model to create a vast range of realistic substitution possibilities that fit the agro-ecological conditions, whereas the fit between expert knowledge or site-specific data from well-managed trails and actual farming practices can sometimes be unclear. The simulation period over which these interventions are assessed is deliberately kept brief, so ignoring the interaction of farm-level decision-making and ecological processes, or environmental spill-overs such as pesticide resistance build-up and changes in natural pest control, is less pertinent than when dealing with an extended simulation period. In terms of temporal analysis, the focus of this study is mainly on the diffusion of IPM in relation to different adoption incentives, for which a short simulation period is sufficient. The acceptance and also the impact of an innovation should become noticeable over a five year horizon if the problem at stake is pressing and policy recommendations imperative.

As well as for running policy-oriented scenarios, the MPMAS application is also a valuable tool for assessing the impacts of science-oriented scenarios. An integrated approach combining SWAT, a model of physico-chemical processes, and MPMAS, a farm-level decision-making model, illustrates the trade-offs between farm incomes and strict environmental sustainability requirements. It also complements policy scenarios in terms of exploring the effect of potential regulations based on such requirements rather than economic instruments. This part of the assessment was built on the premise that daily concentrations should never exceed scientifically determined threshold values. Such a restrictive condition was used, because the simulated concentrations were derived from the daily loads simulated by SWAT using the daily discharge. Therefore, sub-daily concentrations could indeed exceed the thresholds, even if no mean daily concentrations did. To minimize this risk, a very strict criterion was selected. The choice of pesticides for assessing the effect of adhering to the eco-toxicological thresholds was determined by their importance to farmers in Thailand, the availability of data and the SWAT model's capabilities. In the MPMAS model set up, the implementation of a strict scientific criterion evidently engenders a different response from agents than economic (dis-)incentives. Substitution between pesticides is forced within the science-oriented scenarios, while changes in pesticide use are influenced by relative elasticities within the policy-oriented scenarios. This requires that regulations indiscriminately target both high and less profitable production activities. As a result, in the MP with fixed substitution possibilities, the impacts on income levels of regulating important pesticides, such as cypermethrin, are high.

6.3 Implications for pesticide policy-making

First of all, it needs to be stressed that policy implications are related to methodology. Skevas et al. (2012) found that a lack of response by Dutch farmers to increases in the price of pesticides is critically influenced by their choice of model. As they used an econometric simulation model based on an exponential damage control specification, the impact of the tax is determined by the low output elasticity of pesticide use. Therefore, a 120% tax only reduced pesticide applications by 4%. The high values of pesticide overuse that result from the econometric pesticide productivity analysis carried out for Thai farmers with a similar methodological set-up, also suggests that the responsiveness of farmers to higher pesticide tax levels of is very low. From their review of 17 studies and an analysis of the potential for pesticide taxation in Mexico, Pina and Forcada (2004) ascertained that own-price elasticities of pesticide demand are generally low, which leads them to conclude that farmers will not change their on-farm practices, but instead decide to absorb impacts through their incomes. The representative MP farm model of Falconer and Hodge (2000, 2001) showed that only high taxes can achieve significant pesticide use reductions. This is in line with the results of the model used here, which demonstrates that, while the tax impact is far from great, a 50% tax can reduce pesticide use by 8.5%. While the model parameterisation is influenced by the exponential damage control term, the dynamic optimization process is based on a vast array of activities, constraints and pesticide substitution options, and thus provides a more detailed representation of the actual farm household decision-making environment. In conclusion, MP-based farm models seem to yield more perceptible tax impacts than econometric analyses, especially if based on an exponential damage control function. At first glance, the sector MP model of Jacquet et al. (2011) produced extra-ordinary results, since a 16% tax reduced pesticide use by 20%. It must be said however that, in contrast to MPMAS, the model was not calibrated to reproduce observed behaviour, but set up to explore the capacities of a series of low-input technologies without access constraints. In the MPMAS simulation scenario, in which a 50% tax is combined with low-input IPM access, the innovation becomes available to the first three innovator segments and the reduction of pesticide use rises to 13.5%. The availability of technologies and the sequence of access, both of which are a function of individual characteristics and time, influenced the results significantly. Also, the redistribution effect of the tax revenues has an essential impact on the model outcomes. In this regard, the results here are in line with those of Jacquet et al. (2011), who tested the direct compensation of farmers against the provision of subsidies to organic farming. Their model and the MPMAS application used here provide evidence that financially supporting sustainable farming technologies are more effective at achieving the environmental goal of lower pesticide applications, than the per hectare redistribution of collected taxes. The simulation results of both models indicate that, due to the good rate of return provided by low-input innovation, incomes are not negatively affected by policy-induced land-use changes.

For policy makers, it is important of know which policy mix works best. As explained in more detail in Section 5.4, compared to the transitional area payments and the price premium, as well as various mixes, the high level bio-pesticide subsidy is the most cost effective, practicable and realisable adoption incentive policy. The MPMAS model demonstrates that the combination of a high proportional tax and a 60% bio-pesticide subsidy, one financed by tax revenues, reduces overall pesticide use after five simulation periods by 18%, and more toxic pesticide use by almost 25%. On the whole, reductions of that magnitude can be considered significant over a short time horizon. Taxes, in combination with IPM access and a subsidy, have been have proved to deliver substantially higher impacts on pesticide use than the stand-alone introduction of a tax instrument or of IPM. This finding is in line with those of Falconer and Hodge (Falconer, 2000), who stated that taxes can be more effective if farmers are provided with pest control alternatives. The lack of alternatives to synthetic pesticides among Thai farmers has been described as one of the main factors causing the high levels of pesticide use in the country (Lamers et al., 2013; Schreinemachers et al., 2011). Praneetvatakul et al. (2013) concluded that it is best to introduce a package of policies that combines an environmental tax with supportive measures to help farmers change their on-farm practices. This research has confirmed this, by showing that the availability of IPM for one group of vegetable crops can bring down pesticide pollution levels, especially if accompanied by policies to incentivise adoption. Due to a lack of data, IPM methods are only included in the model for leafy vegetables. With integrated pest control available for other cropping activities, in particular bell peppers, tomatoes and flowers, even more far-reaching reductions can be achieved. The same is likely to be the case for a time horizon above five simulation periods; however, due to the innovation diffusion over time, the long-term financing of bio-pesticide subsidies or price premiums might not be covered by tax revenues. The advantage of transitional area payments is that while they are expensive in the short run, they are eventually phased out. Even though they might be unworkable as a widely introduced policy in the northern Thai context, they might present a viable policy option for locally limited project-based pesticide use reduction interventions. One way to guarantee financing for subsidies and price premiums would be to cap the number of eligible farmers or the amount of eligible land. The simulation results show clearly that providing one incentive instrument – input subsidies – at a high level, is more effective than mixing different incentive policies, each of which is at lower levels. With additional funds, especially if resources can be shifted from general subsidies for conventional agriculture to targeted financing of integrated agriculture, it could be possible to sustain a comprehensive (up to 80%) bio-pesticide subsidy and, as a result, increase toxic pesticide use reductions. By period five of the simulation, reductions amount to almost 35%. Transaction costs, for which no estimates were available in the model, can be assumed to be lower when implementing a single incentive policy with a policy infrastructure already in place, rather than a mix of payments, subsidies and price premiums. At the moment however, subsidies for conventional agriculture continue to dominate, and there is a lack of government will to promote products produced by sustainable agriculture in the domestic market, despite official commitments to the contrary. Kasem and Thapa (Kasem and Thapa, 2012) pointed out that there is a clear divergence between the commitments and plans of policy-makers, and actual practice in Thailand.

In addition, a crucial factor in the impact of policies on pesticide use is the nature of the accompanying innovation. GAP certification is not the subject of innovation diffusion in the MPMAS model, as evidence suggests that the public GAP standard in Thailand has little impact on reducing pesticide use due to a lack of standard compliance and changes in on-farm practices (Amekawa, 2013; Schreinemachers et al., 2012). In terms of the MPMAS model's assessment of the impact of pesticide use reduction strategies, it was decided to use a more far-reaching sustainability innovation, that is the IPM system practiced by farmers belonging to Royal Project station in Doi Angkhang. The transfer of data from that location to the Mae Sa watershed can be regarded as unproblematic in terms of crop suitability, since the climate, soils and topography are similar in both locations. Furthermore, using the Doi Angkhang data was predicated on the assumption that a comparable Royal Project infrastructure is available in the watershed, one which supports farmers through the provision of extension and marketing services. While the Royal Project has a strong presence in the study area, extension capacities for IPM promotion need to be further developed. In places, with a lack of solid advisory structures provided by the Royal Project or other extension services, an innovation, such as comprehensive IPM farming, which requires a high degree of skill transfer and logistics to be in place, is more difficult to become established. The major sources of information for most upland farmers are district offices of the Department of Agriculture (DoA), neighbours and shopkeepers. The preparation of biological pest control and fertilising inputs, as well as the sequence of input applications for IPM is knowledge-intensive. In addition, investment is required to improve the capacity of government agencies and NGOs to better respond to farmers' demands, facilitate knowledge exchange and strengthen community initiatives. Farmer field schools have been successfully tested in Southeast Asia in order to implement community IPM (Pontius et al., 2002), but they need to be tailored to the local circumstances. In the context of northern Thailand, IPM has so far played a marginal role. The example of Doi Angkhang, as well as the econometric analysis of GAP certification adoption provided here, suggests that respected institutions that provide effective extension services and also access to high-value markets, are important pre-conditions for the successful introduction and widespread adoption of sustainability innovations. This however does not rule out the introduction of participatory approaches that take into account the diversity of demand for innovations, and allow farmers to adapt innovations to their needs, since their role has been found to be important (Schreinemachers et al., 2013). However, formal advisory services need to be linked to informal farmer networks in such cases. In this study, exposure, that is the links between farmers who are aware and those who are not aware of an innovation, is also shown to be vital for innovation adoption. For public investments in agricultural research aimed at sustainable land use and rural development in the mountainous areas of Southeast Asia, this means policy-makers needs to move from the financing of piecemeal research and technology developments, to the building of long-term interactive knowledge partnerships and innovation networks (Neef et al., 2013).

The outcome of the MPMAS model imply that if the institutional context allows for knowledge of IPM innovations to take root among local communities, the integrated management of pests will be widely adopted and implemented; its profitability also guarantees that the innovation will be sustained in the long run. An important policy conclusion of research conducted by the Uplands Program in the mountainous areas of Southeast Asia, is that different types of interventions need to be balanced. Policies aimed at addressing the environmental externalities caused by market participation for example should be combined with socially-oriented policies that target poorer segments of the population (Zeller et al., 2013). The prediction of innovativeness and the simulation results regarding IPM adoption shown here demonstrate that poorer households can be innovative and so benefit from the introduction of farm-income increasing innovations, such as IPM. Only the wealthier and highly polluting households benefit little. In a nutshell, integrated pest management can partly mitigate the environmental impacts of agricultural intensification, without negatively affecting livelihoods.

Unlike the IPM policy scenarios, the trade-off between strict environmental pesticide use reduction requirements and income levels shown in this study is substantial. To always stay under the stipulated NOEC and PNEC thresholds, the use of cypermethrin has to be reduced by 97.5% and 99.6% respectively, meaning that the current dosage is 250 times above the PNEC threshold and 39 times above the NOEC threshold. For chlorothalonil, the situation is less serious, as the current application rate is below the NOEC threshold, but is still five times above the more restrictive PNEC threshold. Agents' incomes are significantly affected by reductions in either pesticide. Reaching the PNEC threshold for chlorothalonil entails that average incomes fall by 17.3% from the baseline, while for cypermethrin the income drop is even 38.3%. The strict eco-toxicological thresholds imposed produce drastic negative income effects. It could be argued that these thresholds are unrealistic, as farmers could never fulfil such standards under present circumstances. The combination of IPM introduction with economic policy instruments is socially more acceptable and more practicable to implement, it is not possible to guarantee that scientific requirements regarding environmental protection can be sufficiently met though. For this, the adoption of IPM and economic instruments could be complemented by regulations or bans on the use of certain pesticides.

The best approach to reconcile compliance with eco-toxicological thresholds and the economically incentivised transition to integrated pest management needs to be clarified by further research. In this regard, a wide range of integrated pest control methods for all major crops should be assessed. This study has contributed to the pesticide policy debate by clarifying that a trade-off between environmental protection and livelihoods can be avoided if policy-makers manage to introduce economic incentives to motivate changes in growing practices as well as create an enabling environment in which farmers can learn about IPM. In this case, both the environment and farmers benefit.

References

Abadi Ghadim, A.K., Pannell, D.J., 1999. A conceptual framework of adoption of an agricultural innovation. Agricultural Economics 21, 145–154.

Agne, S., 2000. The Impact of Pesticide Taxation on Pesticide Use and Income in Costa Rica's Coffee Production, in: Waibel, H. (Ed.), Pesticide Policy Project Publication Series No. 2. Institute of Horticultural Economics, Hannover University.

Allen, R.G., 1998. Crop evapotranspiration: guidelines for computing crop water requirements, Irrigation and Drainage Series No. 56. Food and Agriculture Organization of the United Nations, Rome.

Amekawa, Y., 2009. Reflections on the Growing Influence of Good Agricultural Practices in the Global South. Journal of Agricultural and Environmental Ethics 22, 531–557.

Amekawa, Y., 2013. Can a public GAP approach ensure safety and fairness? A comparative study of Q-GAP in Thailand. The Journal of Peasant Studies 40, 189–217.

An, L., 2012. Modeling human decisions in coupled human and natural systems: Review of agent-based models. Ecological Modelling 229, 25–36.

Areekul, S., 2007. Insect Pests and Insect Control in Thailand, in: Kunstadter, P. (Ed.), Pesticides in Southeast Asia: Environmental, Biomedical, and Economic Uses and Effects. University of Washington Press, pp. 13–34.

Athisook, K., Sungwaranon, B., Lertreungdej, Y., Jongmeevasana, P., Payanant, T., Chaipolngam, R., 2007. Pesticide residue in foods monitored in Thailand 1999–2003, in: Kunstadter, P. (Ed.), Pesticides in Southeast Asia: Environmental, biomedical, and economic uses and effects. Silkworm Books, Chiang Mai, pp. 151–159.

Atreya, K., 2008. Health costs from short-term exposure to pesticides in Nepal. Social Science & Medicine 67, 511–519.

Bannwarth, M., Sangchan, W., Hugenschmidt, C., Lamers, M., Ingwersen, J., Ziegler, A.D., Streck, T., subm. Pesticide transport simulation in a tropical catchment by SWAT using the ANSELM calibration tool. Journal of Environmental Pollution.

Becu, N., Neef, A., Schreinemachers, P., Sangkapitux, C., 2008. Participatory computer simulation to support collective decision-making: Potential and limits of stakeholder involvement. Land Use Policy 25, 498–509.

Becu, N., Perez, P., Walker, A., Barreteau, O., Page, C.L., 2003. Agent based simulation of a small catchment water management in northern Thailand: Description of the CATCHSCAPE model. Ecological Modelling 170, 319–331.

Berger, T., 2001. Agent-based spatial models applied to agriculture: a simulation tool for technology diffusion, resource use changes and policy analysis. Agricultural Economics 25, 245.

Berger, T., Birner, R., Díaz, J., McCarthy, N., Wittmer, H., 2007. Capturing the complexity of water uses and water users within a multi-agent framework Water Resources Management 21, 129–148.
Berger, T., Schreinemachers, P., 2006. Creating agents and landscapes for multiagent systems from random samples. Ecology and Society 11, 10–19.
Berger, T., Schreinemachers, P., Woelcke, J., 2006. Multi-agent simulation for the targeting of development policies in less-favored areas. Agricultural Systems 88, 28–43.
Buchanan, J., Tullock, G., 1975. Polluters' Profits and Political Response: Direct Controls versus Taxes. The American Economic Review 65, 139–147.
Carpentier, A., Weaver, R.D., 1997. Damage Control Productivity: Why Econometrics Matters. American Journal of Agricultural Economics 79, 47–61.
Carrasco-Tauber, C., Moffitt, L.J., 1992. Damage Control Econometrics: Functional Specification and Pesticide Productivity. American Journal of Agricultural Economics 74, 158–162.
Carvalho, F.P., 2006. Agriculture, pesticides, food security and food safety. Environmental Science & Policy 9, 685–692.
Chambers, R.G., Lichtenberg, E., 1994. Simple Econometrics of Pesticide Productivity. American Journal of Agricultural Economics 76, 407–417.
Clarke, D., Smith, M., El_Askari, K., 1998. New software for crop water requirements and irrigation scheduling. Irrigation and Drainage 47, 45–58.
Doorenbos, J., Kassam, A.H., Bentvelsen, C.I.M., 1979. Yield response to water, Irrigation and Drainage Series No. 33. Food and Agriculture Organization of the United Nations, Rome.
Ecobichon, D.J., 2001. Pesticide use in developing countries. Toxicology 160, 27–33.
Falconer, K., Hodge, I., 2000. Using economic incentives for pesticide usage reductions: responsiveness to input taxation and agricultural systems. Agricultural Systems 63, 175–194.
Falconer, K., Hodge, I., 2001. Pesticide taxation and multi-objective policy-making: farm modelling to evaluate profit:environment trade-offs. Ecological Economics 36, 263–279.
FAO, 2002. International Code of Conduct on the Distribution and Use of Pesticides. Food and Agriculture Organization of the United Nations, Rome.
FAO, 2010. Guidance on Pest and Pesticide Management Policy Development. Food and Agriculture Organization of the United Nations.
FAO, 2011. Save and Grow: A Policymaker's Guide to the Sustainable Intensification of Smallholder Crop Production. FAO, Rome.
FAO, 2013a. FAO CropWat software, 8.0 ed. Land and Water Development Division, FAO, Rome.
FAO, 2013b. FAO Regional Vegetable IPM Programme in Asia – National IPM Programme Thailand. FAO, Bangkok.

Felder, S., Schleiniger, R., 2002. Environmental tax reform: efficiency and political feasibility. Ecological Economics 42, 107–116.

Filatova, T., Verburg, P.H., Parker, D.C., Stannard, C.A., 2013. Spatial agent-based models for socio-ecological systems: Challenges and prospects. Environmental Modelling & Software 45, 1–7.

Giovannucci, D., 2006. Evaluation of Organic Agriculture and Poverty Reduction in Asia. IFAD, Rome.

Granovetter, M., Chung, C., 1997. Social Networks, Organizational Politics and Governance Coalitions: The Relationship Between Industrial Associations and the American Electricity Industry: 1885–1910. Sun Yat-Sen Management Review 6.

Grovermann, C., Schreinemachers, P., Berger, T., 2013. Quantifying pesticide overuse from farmer and societal points of view: An application to Thailand. Crop Protection 53, 161–168.

Halberg, N., Sillebak, K., 1997. Expected Crop Yield Loss When Converting to Organic Dairy Farming in Denmark. Biological Agriculture and Horticulture 14, 25–41.

Hazell, P., 2009. The Asian Green Revolution, IFPRI Discussion Paper 0911. International Food Policy Researcjh Institute (IFPRI), Washington.

Heidhues, F., Pape, J., 2007. Sustainable Land Use and Sustainable Rural Livelihoods in Mountainous Regions of Southeast Asia – Meeting the Challanges of Ecological, Socio-Economic and Cultural Diversity, in: Heidhues, F., Herrmann, L., Neef, A., Neidhart, S., Pape, J., Sruamsiri, P., Thu, D.C., Valle Zárate, A. (Eds.), Sustainable Land Use in Mountainous Regions of Southeast Asia. Springer, Heidelberg.

Horna, D., Smale, M., Al-Hassan, R., Falck-Zepeda, J., Timpo, S.E., 2008. Insecticide use on vegetables in Ghana, in: Institute, I.F.P.R. (Ed.), IFPRI Discussion Paper, Washington.

Huang, J., Hu, R., Rozelle, S., Qiao, F., Pray, C.E., 2002. Transgenic varieties and productivity of smallholder cotton farmers in China. Australian Journal of Agricultural and Resource Economics 46, 367–387.

Jacquet, F., Butault, J.-P., Guichard, L., 2011. An economic analysis of the possibility of reducing pesticides in French field crops. Ecological Economics 70, 1638–1648.

Jaeger, W.K., 2005. Environmental economics for tree huggers and other skeptics. Island Press, London.

Jah, R.K., Regmi, A.P., 2009. Productivity of pesticides in vegetable farming in Nepal, SADEE Working Paper. South Asian Network for Development and Environmental Economics, Katmandu, Nepal.

Janssen, S., van Ittersum, M.K., 2007. Assessing farm innovations and responses to policies: A review of bio-economic farm models. Agricultural Systems 94, 622–636.

Jayamangkala, N., 2006. Vegetable supply chain management: The Royal Project Foundation in Thailand. Acta Horticulturae 699, 43–46.
Jungbluth, F., 1996. Crop Protection Policy in Thailand: Economic and Political Factors Influencing Pesticide Use, in: Waibel, H. (Ed.), Pesticide Policy Project Publication Series No. 6. Institute of Development and Agricultural Economics, University of Hannover.
Kaiser, H.M., Messer, K.D., 2012. Mathematical Programming for Agricultural, Environmental, and Resource Economics. Wiley.
Kasem, S., Thapa, G.B., 2012. Sustainable development policies and achievements in the context of the agriculture sector in Thailand. Sustainable Development 20, 98–114.
Kenmore, P.E., 1991. Indonesia's integrated pest management: A model for Asia. FAO Inter-Country Programme for Integrated Pest Control in South and Southeast Asia.
Konradsen, F., van der Hoek, W., Cole, D.C., Hutchinson, G., Daisley, H., Singh, S., Eddleston, M., 2003. Reducing acute poisoning in developing countries—options for restricting the availability of pesticides. Toxicology 192, 249–261.
Kovach, J., Petzoldt, C., Degni, J., Tette, J., 1992. A method to measure the environmental impact of pesticides. New York's Food and Life Science Bulletin 139, New York Agricultural Experiment Station, Cornell University, Ithaca, NY, 8 pp.
Kunstadter, P., 2007. Introduction, in: Kunstadter, P. (Ed.), Pesticides in Southeast Asia: Environmental, biomedical, and economic uses and effects. Silkworm Books, Chiang Mai, pp. 1–12.
Kuosmanen, T., Pemsl, D., Wesseler, J., 2006. Specification and Estimation of Production Functions Involving Damage Control Inputs: A Two-Stage, Semiparametric Approach. American Journal of Agricultural Economics 88, 499–511.
Lamers, M., Schreinemachers, P., Ingwersen, J., Sangchan, W., Grovermann, C., Berger, T., 2013. Agricultural Pesticide Use in Mountainous Areas of Thailand and Vietnam: Towards Reducing Exposure and Rationalizing Use, in: Fröhlich, H.L., Schreinemachers, P., Stahr, K., Clemens, G. (Eds.), Sustainable Land Use and Rural Development in Southeast Asia: Innovations and Policies for Mountainous Areas. Springer Berlin Heidelberg, pp. 149–173.
Leach, A.W., Mumford, J.D., 2008. Pesticide Environmental Accounting: A method for assessing the external costs of individual pesticide applications. Environmental Pollution 151, 139–147.
Leach, A.W., Mumford, J.D., 2011. Pesticide environmental accounting: a decision-making tool estimating external costs of pesticides. Journal für Verbraucherschutz und Lebensmittelsicherheit 6, 21–26.
Lichtenberg, E., Zilberman, D., 1986. The Econometrics of Damage Control: Why Specification Matters. American Journal of Agricultural Economics 68, 261–273.

Lindner, R.K., 1987. Adoption and diffusion of technology: an overview, in: Champ, B., Highly, E., Remenyi, J. (Eds.), Technological Change in Postharvest Handling and Transportation of Grains in the Humid Tropics. ACIAR Proceedings No. 19. Australian Centre for International Agricultural Research, Canberra, pp. 144–151.

Marohn, C., Schreinemachers, P., Quang, D.V., Berger, T., Siripalangkanont, P., Nguyen, T.T., Cadisch, G., 2013. A software coupling approach to assess low-cost soil conservation strategies for highland agriculture in Vietnam. Environmental Modelling & Software 45, 116–128.

Mundlak, Y., Larson, D., Butzer, R., 1997. The determinants of agricultural production: a cross-country analysis, Policy Research Working Paper Vol. 1. The World Bank.

Neef, A., Ekasingh, B., Friederichsen, R., Becu, N., Lippe, M., Sangkapitux, C., Frör, O., Punyawadee, V., Schad, I., Williams, P., Schreinemachers, P., Neubert, D., Heidhues, F., Cadisch, G., The Dang, N., Gypmantasiri, P., Hoffmann, V., 2013. Participatory Approaches to Research and Development in the Southeast Asian Uplands: Potential and Challenges, in: Fröhlich, H.L., Schreinemachers, P., Stahr, K., Clemens, G. (Eds.), Sustainable Land Use and Rural Development in Southeast Asia: Innovations and Policies for Mountainous Areas. Springer Berlin Heidelberg, pp. 321–365.

Neupane, R.P., Sharma, K.R., Thapa, G.B., 2002. Adoption of agroforestry in the hills of Nepal: a logistic regression analysis. Agricultural Systems 72, 177–196.

Nolan, J., Parker, D., Van Kooten, G.C., Berger, T., 2009. An Overview of Computational Modeling in Agricultural and Resource Economics. Canadian Journal of Agricultural Economics/Revue canadienne d'agroeconomie 57, 417–429.

Oskam, A.J., 1997. The economics of pesticides: An overview of the issues, in: Oskam, A.J., Vijftigschild, T.A.M. (eds.) (Ed.), Proceedings and discussions of the workshop on pesticides, Wageningen, pp. 360–384.

Parker, D.C., Manson, S. M., Janssen, M. A., Hoffmann, M. J., Deadman, P., 2003. Multi-Agent System Models for the Simulation of Land-Use and Land-Cover Change: A Review. Annals of the Association of American Geographers 93.

Pemsl, D., Waibel, H., Gutierrez, A.P., 2005. Why Do Some Bt-Cotton Farmers in China Continue to Use High Levels of Pesticides? International Journal of Agricultural Sustainability 3, 44–56.

Pimentel, D., 2005. 'Environmental and Economic Costs of the Application of Pesticides Primarily in the United States'. Environment, Development and Sustainability 7, 229–252.

Pimentel, D., McLaughlin, L., Zepp, A., Lakitan, B., Kraus, T., Kleinman, P., Vancini, F., Roach, W.J., Graap, E., Keeton, W.S., Selig, G., 1993. Environmental and economic effects of reducing pesticide use in agriculture. Agriculture, Ecosystems & Environment 46, 273–288.

Pina, C.M., Forcada, S.A., 2004. Effects of an environmental tax on pesticides in Mexico. Industry and Environment 27, 34–38.
Pingali, P.L., 2001. Environmental consequences of agricultural commercialization in Asia. Environment and Development Economics 6, 483–502.
Pingali, P.L., Rosegrant, M.W., 1995. Agricultural commercialization and diversification: processes and policies. Food Policy 20, 171–185.
Pontius, J., Dilts, R., Bartlett, A., 2002. Ten Years of IPM Training in Asia – From Farmer Field School to Community IPM. FAO, Bangkok.
Praneetvatakul, S., 2005. Pesticides policies in Thailand Social Research Institute, Chiang Mai University.
Praneetvatakul, S., Kuwattanasiri, S.D., Waibel, H., 2003. The Productivity of Pesticide use in rice production of Thailand: A damage control approach. Thai Journal of Agricultural Economics 22.
Praneetvatakul, S., Schreinemachers, P., Pananurak, P., Tipraqsa, P., 2013. Pesticides, external costs and policy options for Thai agriculture. Environmental Science & Policy 27, 103–113.
Praneetvatakul, S., Waibel, H., Meenakanit, L., 2007. Farmer Field Schools in Thailand: History, Economics and Policy, in: Waibel, H. (Ed.), Pesticide Policy Project Publication Series No. 12. Institute of Development and Agricultural Economics, Leibniz University of Hannover.
Pretty, J., Brett, C., Gee, D., Hine, R., Mason, C., Morison, J., Raven, H., Rayment, M., van der Bijl, G., 2000. An assessment of the total external costs of UK agriculture. Agricultural Systems 65, 113–136.
Pretty, J., Brett, C., Gee, D., Hine, R., Mason, C., Morison, J., Rayment, M., Bijl, G.V.D., Dobbs, T., 2001. Policy and practice: policy challenges and priorities for internalizing the externalities of modern agriculture. Journal of Environmental Planning and Management 44, 263–283.
Qaim, M., De Janvry, A., 2005. Bt cotton and pesticide use in Argentina: economic and environmental effects. Environment and Development Economics 10, 179–200.
Quang, D.V., Schreinemachers, P., Berger, T., 2014. Ex-ante assessment of soil conservation methods in the uplands of Vietnam: An agent-based modeling approach. Agricultural Systems, 108–119.
Rogers, E.M., 2003. Diffusion of Innovations, 5 ed. Simon and Schuster.
Sangchan, W., Bannwarth, M., Ingwersen, J., Hugenschmidt, C., Schwadorf, K., Thavornyutikarn, P., Pansombat, K., Streck, T., 2013. Monitoring and risk assessment of pesticides in a tropical river of an agricultural watershed in northern Thailand. Environmental Monitoring and Assessment, 1–17.
Sangchan, W., Hugenschmidt, C., Ingwersen, J., Schwadorf, K., Thavornyutikarn, P., Pansombat, K., Streck, T., 2012. Short-term dynamics of pesticide concentrations and loads in a river of an agricultural watershed in the outer tropics. Agriculture, Ecosystems & Environment 158, 1–14.

Sathāban Wičhai, Phatthanā Phư̄nthī Sūng, Khrōngkān Lūang, 2007. The peach and the poppy: the story of Thailand's Royal Project. Highland Research and Development Institute, Chiang Mai, Thailand.

Schipmann, C., Qaim, M., 2010. Spillovers from modern supply chains to traditional markets: product innovation and adoption by smallholders. Agricultural Economics 41, 361–371.

Schou, J.S., Streibig, J. C., 1999. Pesticide Taxes in Scandinavia. Pesticide Outlook 10, 127–129.

Schreinemachers, P., Berger, T., 2011. An agent-based simulation model of human–environment interactions in agricultural systems. Environmental Modelling & Software 26, 845–859.

Schreinemachers, P., Berger, T., Aune, J.B., 2007a. Simulating soil fertility and poverty dynamics in Uganda: A bio-economic multi-agent systems approach. Ecological Economics 64, 387–401.

Schreinemachers, P., Berger, T., Sirijinda, A., Praneetvatakul, S., 2009. The diffusion of greenhouse agriculture in Northern Thailand: Combining econometrics and Agent-Based Modeling. Canadian Journal of Agricultural Economics/Revue canadienne d'agroeconomie 57, 513–536.

Schreinemachers, P., Berger, T., 2006. Land use decisions in developing countries and their representation in multi-agent systems. Journal of Land Use Science 1, 29–44.

Schreinemachers, P., Fröhlich, H., Clemens, G., Stahr, K., 2013. From Challenges to Sustainable Solutions for Upland Agriculture in Southeast Asia, in: Fröhlich, H.L., Schreinemachers, P., Stahr, K., Clemens, G. (Eds.), Sustainable Land Use and Rural Development in Southeast Asia: Innovations and Policies for Mountainous Areas. Springer Berlin Heidelberg, pp. 3–27.

Schreinemachers, P., Potchanasin, C., Berger, T., Roygrong, S., 2010. Agent-based modeling for ex-ante assessment of tree crop technologies: litchis in northern Thailand. Agricultural Economics 41, 519–536.

Schreinemachers, P., Praneetvatakul, S., Sirijinda, A., Berger, T., 2008. Agricultural statistics of the Mae Sa watershed area, Thailand, 2006. The Uplands Program, Chiang Mai.

Schreinemachers, P., S., Sirijinda, A., Sringam, S., Berger, T., 2010. Agricultural statistics of the Mae Sa watershed area, Thailand, 2009/10. The Uplands Program, Chiang Mai.

Schreinemachers, P., Schad, I., Tipraqsa, P., Williams, P., Neef, A., Riwthong, S., Sangchan, W., Grovermann, C., 2012. Can public GAP standards reduce agricultural pesticide use? The case of fruit and vegetable farming in northern Thailand. Agriculture and Human Values 29, 519–529.

Schreinemachers, P., Sringarm, S., Sirijinda, A., 2011. The role of synthetic pesticides in the intensification of highland agriculture in Thailand. Crop Protection 30, 1430–1437.

Schreinemachers, P., Tipraqsa, P., 2012. Agricultural pesticides and land use intensification in high, middle and low income countries. Food Policy 37, 616–626.
Schreinemachers, P., Tipraqsa, P., Berger, T., 2007b. Assessing innovations and sustainability strategies with multi-agent systems, International Symposium towards Sustainable Livelihoods and Ecosystems in Mountainous Regions, Chiang Mai.
Seufert, V., Ramankutty, N., Foley, J.A., 2012. Comparing the yields of organic and conventional agriculture. Nature 485, 229–232.
Sexton, S., Zhen, L., Zilberman, D., 2007. The Economics of Pesticides and Pest Control. International Review of Environmental and Resource Economics 1, 271–326.
Shankar, B., Thirtle, C., 2005. Pesticide Productivity and Transgenic Cotton Technology: The South African Smallholder Case. Journal of Agricultural Economics 56, 97–116.
Shiferaw, B.A., Kebede, T.A., You, L., 2008. Technology adoption under seed access constraints and the economic impacts of improved pigeonpea varieties in Tanzania. Agricultural Economics 39, 309–323.
Skevas, T., Stefanou, S.E., Lansink, A.O., 2012. Can economic incentives encourage actual reductions in pesticide use and environmental spillovers? Agricultural Economics 43, 267–276.
Snelder, D.J., Masipiqueña, M.D., de Snoo, G.R., 2008. Risk assessment of pesticide usage by smallholder farmers in the Cagayan Valley (Philippines). Crop Protection 27, 747–762.
Thapa, G.B., Rattanasuteerakul, K., 2011. Adoption and extent of organic vegetable farming in Mahasarakham province, Thailand. Applied Geography 31, 201–209.
Thapinta, A., Hudak, P.F., 2000. Pesticide Use and Residual Occurrence in Thailand. Environmental Monitoring and Assessment 60.
Tilman, D., 1999. Global environmental impacts of agricultural expansion: The need for sustainable and efficient practices. Proceedings of the National Academy of Sciences 96, 5995–6000.
Tilman, D., Cassman, K.G., Matson, P.A., Naylor, R., Polasky, S., 2002. Agricultural sustainability and intensive production practices. Nature 418, 671–677.
Tomlin, C.D.S., 2009. The pesticide manual: A World compendium. British Crop Protection Council (11th Revised edition), Alton, U.K.
Tscharntke, T., Klein, A.M., Kruess, A., Steffan-Dewenter, I., Thies, C., 2005. Landscape perspectives on agricultural intensification and biodiversity – ecosystem service management. Ecology Letters 8, 857–874.
University_of_Herdfordshire, 2013. The Pesticide Properties DataBase (PPDB) developed by the Agriculture & Environment Research Unit (AERU), University of Herdfordshire, 2006–2013.

Unnevehr, L.J., 2000. Food safety issues and fresh food product exports from LDCs. Agricultural Economics 23, 231–240.

Valente, T., 2012. Diffusion Processes in Social Networks, in: Meyers, R.A. (Ed.), Computational Complexity. Springer New York, pp. 2940–2952.

Valente, T.W., 2005. Models and methods for innovation diffusion, in: P. J. Carrington, J. Scott, S. Wasserman (Eds.), Models and Methods in Social Network Analysis. Cambridge University Press, Cambridge, UK.

Valente, T.W., Hoffman, B.R., Ritt-Olson, A., Lichtman, K., Johnson, C.A., 2003. Effects of a Social-Network Method for Group Assignment Strategies on Peer-Led Tobacco Prevention Programs in Schools. American Journal of Public Health 93, 1837–1843.

Van de Ven, W.P.M.M., Van Praag, B.M.S., 1981. The demand for deductibles in private health insurance: A probit model with sample selection. Journal of Econometrics 17, 229–252.

Vongsaroj, P., 2007. Economics of IPM, in: Kunstadter, P. (Ed.), Pesticides in Southeast Asia – Environmental, Biomedical, and Economic Uses and Effects. Silkworm Books, Bangkok, pp. 249–259.

Waibel, H., Fleischer, G., Kenmore, P.E., Feder, G., 1998. Evaluation of IPM Programs – Concepts and Methodologies, in: Waibel, H. (Ed.), Pesticide Policy Project Publication Series No. 8. Institute of Development and Agricultural Economics, University of Hannover.

WHO, 2009. The WHO recommended classification of pesticides by hazard and guidelines to classification. World Health Organization, Geneva.

Wilby, A., Thomas, M.B., 2002. Natural enemy diversity and pest control: patterns of pest emergence with agricultural intensification. Ecology Letters 5, 353–360.

Wilson, C., Tisdell, C., 2001. Why farmers continue to use pesticides despite environmental, health and sustainability costs. Ecological Economics 39, 449–462.

Wollni, M., Zeller, M., 2007. Do farmers benefit from participating in specialty markets and cooperatives? The case of coffee marketing in Costa Rica1. Agricultural Economics 37, 243–248.

Wossen, T., Berger, T., Mequaninte, T., Alamirew, B., 2013. Social network effects on the adoption of sustainable natural resource management practices in Ethiopia. International Journal of Sustainable Development & World Ecology 20, 477–483.

Zeller, M., Ufer, S., Van, D., Nielsen, T., Schreinemachers, P., Tipraqsa, P., Berger, T., Saint-Macary, C., Van, L., Keil, A., Dung, P., Heidhues, F., 2013. Policies for Sustainable Development: The Commercialization of Smallholder Agriculture, in: Fröhlich, H.L., Schreinemachers, P., Stahr, K., Clemens, G. (Eds.), Sustainable Land Use and Rural Development in Southeast Asia: Innovations and Policies for Mountainous Areas. Springer Berlin Heidelberg, pp. 463–490.

Annex

Annex I: Percentage of IPM adopters in the agent population

Scenario name	Period 1 Adopters (%)	Period 2 Adopters(%)	Period 3 Adopters(%)	Period 4 Adopters(%)	Period 5 Adopters(%)	Period 5 IPM area (ha)
ALL_____________	0.696	0.846	0.867	0.881	0.893	470.71
IPM_____________	0.020	0.022	0.145	0.418	0.435	207.25
IPMTF1_________	0.022	0.022	0.144	0.428	0.439	210.32
IPMTP1_________	0.021	0.023	0.145	0.426	0.440	210.45
IPMTC1_________	0.021	0.022	0.146	0.431	0.437	210.99
IPMTF2_________	0.022	0.022	0.145	0.421	0.437	213.75
IPMTP2_________	0.021	0.023	0.144	0.426	0.439	213.39
IPMTC2_________	0.021	0.023	0.146	0.426	0.437	215.06
IPMTF3_________	0.022	0.022	0.144	0.420	0.438	213.85
IPMTP3_________	0.021	0.023	0.143	0.425	0.440	216.02
IPMTC3_________	0.021	0.023	0.147	0.427	0.437	215.89
IPM___PA1______	0.021	0.022	0.148	0.427	0.439	219.63
IPM___PA2______	0.021	0.022	0.148	0.434	0.443	230.61
IPM___PA3______	0.022	0.023	0.150	0.449	0.759	421.44
IPM______PR1___	0.021	0.022	0.149	0.429	0.441	218.60
IPM______PR2___	0.021	0.022	0.144	0.435	0.445	230.44
IPM______PR3___	0.021	0.023	0.150	0.449	0.756	413.92
IPM_________SU1	0.021	0.022	0.148	0.426	0.440	216.89
IPM_________SU2	0.021	0.023	0.149	0.426	0.440	224.04

Scenario name	Period 1 Adopters (%)	Period 2 Adopters(%)	Period 3 Adopters(%)	Period 4 Adopters(%)	Period 5 Adopters(%)	Period 5 IPM area (ha)
IPM_________SU3	0.021	0.022	0.146	0.441	0.449	233.16
IPMTP3PA1PR1___	0.022	0.021	0.148	0.431	0.443	235.54
IPMTP3PA1PR2___	0.022	0.023	0.150	0.437	0.447	247.06
IPMTP3PA1___SU2	0.022	0.023	0.150	0.431	0.445	241.62
IPMTP3PA1___SU3	0.022	0.023	0.152	0.449	0.747	417.05
IPMTP3PA1PR1SU1	0.021	0.022	0.144	0.432	0.443	242.61
IPMTP3PA1PR1SU2	0.022	0.023	0.148	0.434	0.445	418.97
IPMTP3PA1PR2SU1	0.022	0.022	0.148	0.441	0.749	441.29
IPMTP3PA1PR2SU2	0.022	0.023	0.152	0.445	0.758	427.90
IPMTP3PA2______	0.022	0.023	0.150	0.445	0.759	236.06
IPMTP3___PR2___	0.022	0.023	0.147	0.432	0.444	237.46
IPMTP3___PR2SU1	0.022	0.023	0.148	0.442	0.447	244.32
IPMTP3___PR2SU2	0.021	0.023	0.150	0.446	0.756	420.79
IPMTP3___PR2SU3	0.021	0.021	0.153	0.448	0.757	433.62
IPM_________SU3	0.022	0.023	0.146	0.433	0.445	239.93

Annex II: Flat tax for 3 selected scenarios + 2 additional scenarios with higher tax rates

Scenario name	Tax rate	Δ income		Tax revenues	Δ pesticide use	Income loss effectiveness	Tax effectiveness
		(%)	(1000 baht/ hh)	(1000 baht/ hh)	(%)	(Δ income/ Δ pesticide use)	(Δ pesticide use/ tax revenues)
TF1	10%	–1.59	–4.10	1.44	–2.62	1.56	–1.82
TF2	30%	–5.12	–13.09	4.10	–5.93	2.21	–1.45
TF3	50%	–7.45	–19.05	6.56	–8.48	2.25	–1.29
TF4	75%	–10.72	–27.35	9.30	–11.17	2.45	–1.20
TF5	100%	–13.94	–35.56	11.90	–13.25	2.68	–1.11

Annex III: Land-use shares in the different scenarios

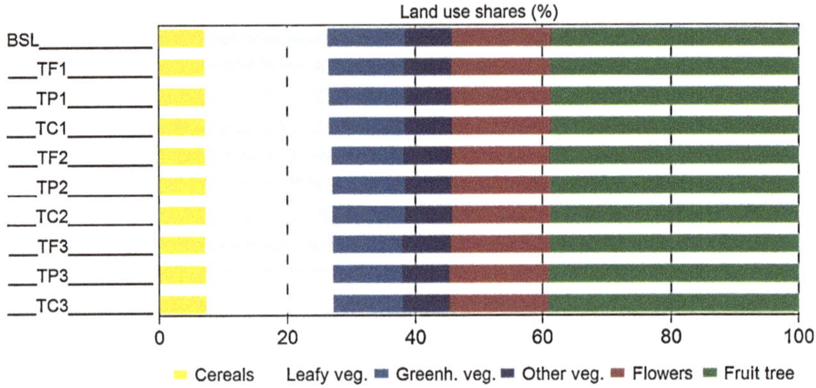

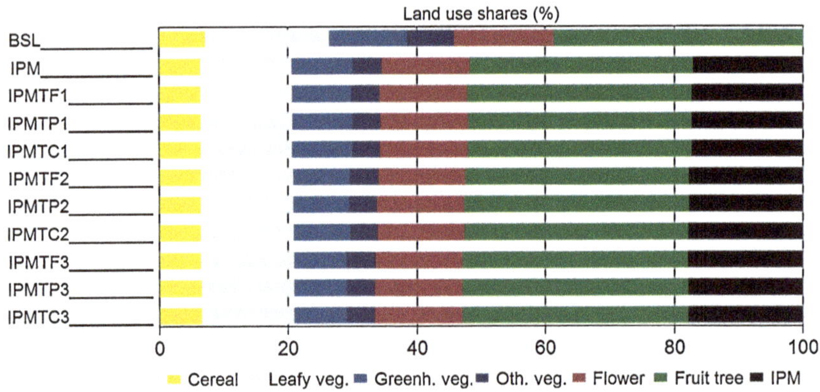

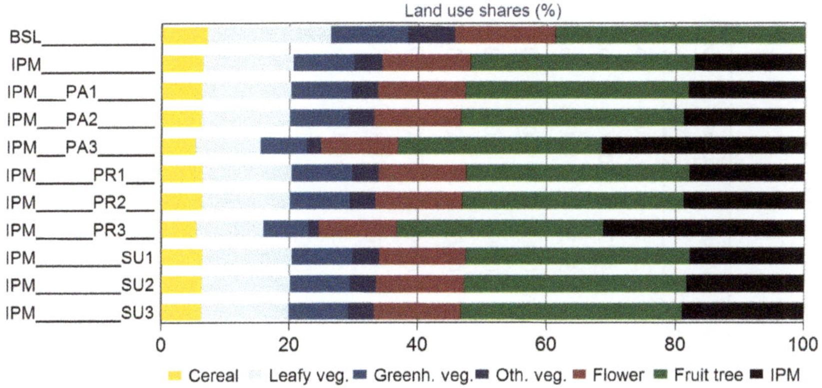

Land use shares (%)
BSL
IPM
IPM___PA1
IPM___PA2
IPM___PA3
IPM______PR1
IPM______PR2
IPM______PR3
IPM_________SU1
IPM_________SU2
IPM_________SU3
0
20
40
60
80
100
Cereal
Leafy veg.
Greenh. veg.
Oth. veg.
Flower
Fruit tree
IPM

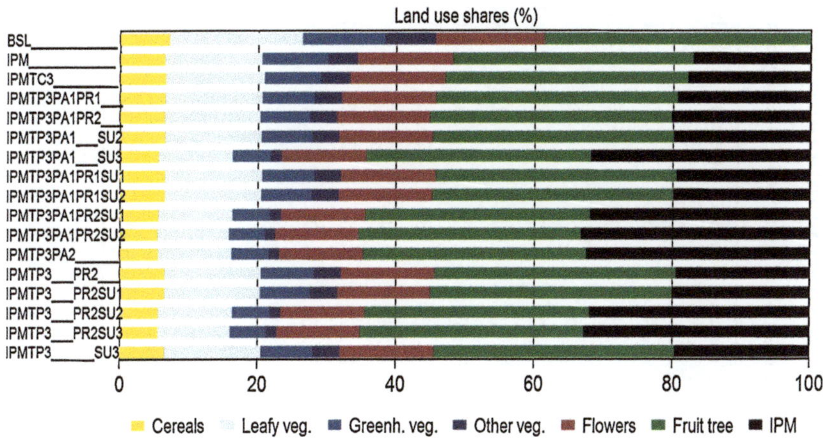

Land use shares (%)
BSL
IPM
IPMTC3
IPMTP3PA1PR1
IPMTP3PA1PR2
IPMTP3PA1___SU2
IPMTP3PA1___SU3
IPMTP3PA1PR1SU1
IPMTP3PA1PR1SU2
IPMTP3PA1PR2SU1
IPMTP3PA1PR2SU2
IPMTP3PA2
IPMTP3___PR2
IPMTP3___PR2SU1
IPMTP3___PR2SU2
IPMTP3___PR2SU3
IPMTP3______SU3
0
20
40
60
80
100
Cereals
Leafy veg.
Greenh. veg.
Other veg.
Flowers
Fruit tree
IPM

Annex IV: Segmented cumulative distribution functions for innovativeness determinants

Difference between individual adoption and first mention of adoption in the study area (years)

Cluster ID	Agents	Cumul (%)	Value	Cumul (%)	Value	Cumul (%)	Value	Cumul (%)	Value	Cumul (%)	Value
0	214	2.22	4.25	4.44	4.50	6.67	5.00	8.89	5.25	100.00	6.00
1	216	4.65	4.00	6.98	4.25	9.30	4.50	13.95	4.75	100.00	5.99
2	171	2.33	2.50	4.65	3.75	6.98	4.25	9.30	5.25	100.00	6.00
3	48	100.00	6.00	0.00	0.00	0.00	0.00	0.00	0.00	0.00	0.00
4	36	16.67	4.75	100.00	6.00	0.00	0.00	0.00	0.00	0.00	0.00
5	42	14.29	3.25	100.00	6.00	0.00	0.00	0.00	0.00	0.00	0.00
6	110	7.69	3.25	15.38	3.88	26.92	4.33	42.31	4.94	100.00	5.97
7	118	9.09	2.83	21.21	3.94	27.27	4.38	39.39	4.88	100.00	5.98
8	147	4.17	2.75	16.67	4.00	33.33	4.75	37.50	5.25	100.00	6.00
9	41	100.00	6.00	0.00	0.00	0.00	0.00	0.00	0.00	0.00	0.00
10	39	100.00	6.00	0.00	0.00	0.00	0.00	0.00	0.00	0.00	0.00
11	35	100.00	6.00	0.00	0.00	0.00	0.00	0.00	0.00	0.00	0.00
12	92	14.29	4.25	21.43	5.75	100.00	6.00	0.00	0.00	0.00	0.00
13	88	23.08	3.75	30.77	4.00	100.00	6.00	0.00	0.00	0.00	0.00
14	94	8.33	0.00	16.67	2.50	25.00	3.50	33.33	3.75	100.00	6.00

Household size (#)

Cluster ID	Agents	Cumul (%)	Value	Cumul (%)	Value	Cumul (%)	Value	Cumul (%)	Value	Cumul (%)	Value
0	214	2.22	1.00	15.56	2.00	42.22	3.00	86.67	4.00	100.00	5.67
1	216	2.33	1.00	18.60	2.00	44.19	3.00	86.05	4.00	100.00	5.17
2	171	6.98	2.00	51.16	3.00	88.37	4.00	95.35	5.00	100.00	6.00
3	48	25.00	2.00	50.00	3.00	75.00	4.00	100.00	5.00	0.00	0.00
4	36	16.67	1.00	33.33	2.00	66.67	3.00	83.33	4.00	100.00	6.00
5	42	28.57	2.00	57.14	4.00	85.71	5.00	100.00	10.00	0.00	0.00
6	110	3.85	2.00	34.62	3.75	80.77	5.42	92.31	7.33	100.00	10.00
7	118	15.15	2.60	45.45	4.50	69.70	6.63	93.94	8.25	100.00	13.50
8	147	20.83	4.60	54.17	6.88	83.33	8.43	91.67	10.50	100.00	14.00
9	41	28.57	2.00	57.14	4.00	100.00	5.00	0.00	0.00	0.00	0.00
10	39	14.29	2.00	28.57	7.00	71.43	8.00	100.00	9.00	0.00	0.00
11	35	28.57	6.00	71.43	7.00	85.71	8.00	100.00	10.00	0.00	0.00
12	92	14.29	3.00	35.71	4.00	64.29	5.00	78.57	6.50	100.00	9.33
13	88	7.69	4.00	15.38	5.00	30.77	6.00	84.62	7.57	100.00	10.50
14	94	16.67	4.00	33.33	6.00	50.00	7.00	83.33	9.00	100.00	16.00

Farm size (ha)

Cluster ID	Agents	Cumul (%)	Value	Cumul (%)	Value	Cumul (%)	Value	Cumul (%)	Value	Cumul (%)	Value
0	214	11.11	0.03	35.56	0.04	62.22	0.06	84.44	0.09	100.00	0.14
1	216	2.33	0.10	16.28	0.11	53.49	0.15	74.42	0.20	100.00	0.28
2	171	2.33	0.16	16.28	0.24	41.86	0.28	72.09	0.42	100.00	0.86
3	48	12.50	0.03	25.00	0.03	50.00	0.05	75.00	0.06	100.00	0.12
4	36	16.67	0.11	33.33	0.12	50.00	0.13	66.67	0.24	100.00	0.68
5	42	14.29	0.16	28.57	0.32	42.86	0.35	71.43	0.47	100.00	0.70
6	110	7.69	0.04	23.08	0.08	42.31	0.10	61.54	0.16	100.00	0.30
7	118	3.03	0.12	15.15	0.17	27.27	0.26	75.76	0.33	100.00	0.66
8	147	12.50	0.27	29.17	0.37	50.00	0.43	66.67	0.48	100.00	0.69
9	41	14.29	0.08	28.57	0.10	42.86	0.12	71.43	0.13	100.00	0.28
10	39	14.29	0.15	28.57	0.16	42.86	0.18	71.43	0.20	100.00	0.41
11	35	14.29	0.35	28.57	0.42	42.86	0.43	71.43	0.50	100.00	0.97
12	92	14.29	0.10	28.57	0.12	42.86	0.14	71.43	0.21	100.00	0.35
13	88	15.38	0.17	30.77	0.23	53.85	0.26	84.62	0.32	100.00	0.46
14	94	16.67	0.25	33.33	0.38	50.00	0.42	66.67	0.46	100.00	0.77

Percentage of high value vegetables (%)

Cluster ID	Agents	Cumul (%)	Value	Cumul (%)	Value	Cumul (%)	Value	Cumul (%)	Value	Cumul (%)	Value
0	214	37.78	0.01	42.22	0.27	51.11	0.48	55.56	0.69	100.00	0.99
1	216	48.84	0.02	58.14	0.24	67.44	0.48	76.74	0.69	100.00	0.97
2	171	41.86	0.00	53.49	0.16	67.44	0.31	76.74	0.66	100.00	0.97
3	48	12.50	0.00	25.00	0.33	50.00	0.57	75.00	0.63	100.00	0.90
4	36	16.67	0.00	33.33	0.43	50.00	0.46	66.67	0.58	100.00	0.70
5	42	14.29	0.27	28.57	0.30	42.86	0.38	71.43	0.47	100.00	0.86
6	110	38.46	0.01	46.15	0.26	61.54	0.47	80.77	0.64	100.00	0.86
7	118	36.36	0.00	48.48	0.12	60.61	0.25	75.76	0.34	100.00	0.50
8	147	29.17	0.01	45.83	0.12	66.67	0.22	83.33	0.30	100.00	0.47
9	41	57.14	0.00	71.43	0.20	85.71	0.50	100.00	1.00	0.00	0.00
10	39	100.00	0.00	0.00	0.00	0.00	0.00	0.00	0.00	0.00	0.00
11	35	85.71	0.00	100.00	0.09	0.00	0.00	0.00	0.00	0.00	0.00
12	92	14.29	0.34	28.57	0.46	42.86	0.55	71.43	0.73	100.00	0.93
13	88	7.69	0.21	23.08	0.50	38.46	0.63	69.23	0.86	100.00	0.93
14	94	16.67	0.49	33.33	0.61	50.00	0.70	66.67	0.73	100.00	0.89

Percentage of litchi (%)

Cluster ID	Agents	Cumul (%)	Value	Cumul (%)	Value	Cumul (%)	Value	Cumul (%)	Value	Cumul (%)	Value
0	214	100.00	0.00	0.00	0.00	0.00	0.00	0.00	0.00	0.00	0.00
1	216	97.67	0.00	100.00	0.40	0.00	0.00	0.00	0.00	0.00	0.00
2	171	93.02	0.00	95.35	0.01	97.67	0.14	100.00	0.14	0.00	0.00
3	48	100.00	0.00	0.00	0.00	0.00	0.00	0.00	0.00	0.00	0.00
4	36	100.00	0.00	0.00	0.00	0.00	0.00	0.00	0.00	0.00	0.00
5	42	71.43	0.00	85.71	0.15	100.00	0.34	0.00	0.00	0.00	0.00
6	110	46.15	0.00	53.85	0.20	69.23	0.31	84.62	0.41	100.00	0.83
7	118	27.27	0.03	42.42	0.24	54.55	0.34	69.70	0.54	100.00	0.77
8	147	16.67	0.10	33.33	0.35	50.00	0.49	66.67	0.63	100.00	0.88
9	41	100.00	0.00	0.00	0.00	0.00	0.00	0.00	0.00	0.00	0.00
10	39	42.86	0.00	57.14	0.26	71.43	0.29	85.71	0.56	100.00	0.63
11	35	14.29	0.00	28.57	0.11	42.86	0.20	71.43	0.31	100.00	0.43
12	92	100.00	0.00	0.00	0.00	0.00	0.00	0.00	0.00	0.00	0.00
13	88	100.00	0.00	0.00	0.00	0.00	0.00	0.00	0.00	0.00	0.00
14	94	83.33	0.00	91.67	0.13	100.00	0.15	0.00	0.00	0.00	0.00

Farm age (years)

Cluster ID	Agents	Cumul (%)	Value	Cumul (%)	Value	Cumul (%)	Value	Cumul (%)	Value	Cumul (%)	Value
0	214	13.33	4.00	24.44	9.40	46.67	17.50	60.00	24.17	100.00	35.33
1	216	9.30	4.25	39.53	12.38	69.77	21.15	88.37	33.88	100.00	40.00
2	171	11.63	4.80	23.26	9.60	34.88	15.60	60.47	22.27	100.00	34.06
3	48	12.50	8.00	25.00	15.00	50.00	20.50	75.00	31.50	100.00	38.00
4	36	16.67	13.00	33.33	19.00	50.00	20.00	66.67	24.00	100.00	27.50
5	42	14.29	15.00	28.57	22.00	42.86	26.00	71.43	34.00	100.00	39.00
6	110	11.54	1.56	34.62	12.00	65.38	18.75	84.62	28.00	100.00	39.50
7	118	3.03	4.00	27.27	13.00	51.52	21.00	69.70	28.00	100.00	43.00
8	147	16.67	9.75	37.50	18.00	58.33	24.60	83.33	31.33	100.00	45.50
9	41	14.29	2.00	28.57	3.00	42.86	9.00	71.43	17.50	100.00	32.50
10	39	14.29	11.00	42.86	20.00	57.14	34.00	85.71	35.00	100.00	40.00
11	35	28.57	15.00	57.14	20.00	71.43	28.00	85.71	30.00	100.00	60.00
12	92	14.29	3.00	28.57	8.50	50.00	16.00	71.43	20.33	100.00	31.25
13	88	15.38	5.00	30.77	12.00	69.23	19.60	84.62	23.50	100.00	36.00
14	94	16.67	9.50	33.33	18.50	50.00	23.00	83.33	28.50	100.00	36.50

Age household head (years)

Cluster ID	Agents	Cumul (%)	Value	Cumul (%)	Value	Cumul (%)	Value	Cumul (%)	Value	Cumul (%)	Value
0	214	6.67	27.00	15.56	37.00	28.89	42.83	71.11	50.42	100.00	61.31
1	216	11.63	33.80	30.23	39.50	41.86	44.80	65.12	49.80	100.00	56.53
2	171	4.65	32.50	18.60	39.83	44.19	45.36	62.79	51.50	100.00	61.50
3	48	12.50	42.00	25.00	48.00	37.50	50.00	62.50	53.50	100.00	59.00
4	36	16.67	31.00	33.33	43.00	50.00	49.00	66.67	53.00	100.00	58.00
5	42	14.29	39.00	28.57	41.00	42.86	46.00	71.43	53.00	100.00	57.50
6	110	11.54	28.67	34.62	35.33	57.69	42.50	80.77	49.67	100.00	55.20
7	118	9.09	28.67	33.33	38.13	57.58	44.75	75.76	50.33	100.00	62.13
8	147	20.83	36.40	41.67	42.20	66.67	47.83	83.33	52.75	100.00	66.00
9	41	14.29	22.00	28.57	28.00	42.86	43.00	57.14	44.00	100.00	55.00
10	39	14.29	28.00	28.57	37.00	42.86	39.00	71.43	47.50	100.00	53.00
11	35	14.29	36.00	28.57	37.00	42.86	38.00	71.43	41.50	100.00	53.50
12	92	21.43	25.67	35.71	32.50	50.00	35.50	64.29	38.50	100.00	48.20
13	88	7.69	27.00	23.08	30.50	38.46	34.50	69.23	40.75	100.00	51.25
14	94	8.33	29.00	25.00	31.00	41.67	38.50	66.67	44.33	100.00	53.75

Education (yes = 1)

Cluster ID	Agents	Cumul (%)	Value	Cumul (%)	Value	Cumul (%)	Value	Cumul (%)	Value	Cumul (%)	Value
0	214	6.67	0.00	100.00	1.00	0.00	0.00	0.00	0.00	0.00	0.00
1	216	100.00	1.00	0.00	0.00	0.00	0.00	0.00	0.00	0.00	0.00
2	171	6.98	0.00	100.00	1.00	0.00	0.00	0.00	0.00	0.00	0.00
3	48	100.00	1.00	0.00	0.00	0.00	0.00	0.00	0.00	0.00	0.00
4	36	100.00	1.00	0.00	0.00	0.00	0.00	0.00	0.00	0.00	0.00
5	42	100.00	1.00	0.00	0.00	0.00	0.00	0.00	0.00	0.00	0.00
6	110	26.92	0.00	100.00	1.00	0.00	0.00	0.00	0.00	0.00	0.00
7	118	45.45	0.00	100.00	1.00	0.00	0.00	0.00	0.00	0.00	0.00
8	147	37.50	0.00	100.00	1.00	0.00	0.00	0.00	0.00	0.00	0.00
9	41	28.57	0.00	100.00	1.00	0.00	0.00	0.00	0.00	0.00	0.00
10	39	71.43	0.00	100.00	1.00	0.00	0.00	0.00	0.00	0.00	0.00
11	35	14.29	0.00	100.00	1.00	0.00	0.00	0.00	0.00	0.00	0.00
12	92	42.86	0.00	100.00	1.00	0.00	0.00	0.00	0.00	0.00	0.00
13	88	23.08	0.00	100.00	1.00	0.00	0.00	0.00	0.00	0.00	0.00
14	94	25.00	0.00	100.00	1.00	0.00	0.00	0.00	0.00	0.00	0.00

Grow more than 1 crop (yes = 1)

Cluster ID	Agents	Cumul (%)	Value	Cumul (%)	Value	Cumul (%)	Value	Cumul (%)	Value	Cumul (%)	Value
0	214	55.56	0.00	100.00	1.00	0.00	0.00	0.00	0.00	0.00	0.00
1	216	27.91	0.00	100.00	1.00	0.00	0.00	0.00	0.00	0.00	0.00
2	171	39.53	0.00	100.00	1.00	0.00	0.00	0.00	0.00	0.00	0.00
3	48	12.50	0.00	100.00	1.00	0.00	0.00	0.00	0.00	0.00	0.00
4	36	16.67	0.00	100.00	1.00	0.00	0.00	0.00	0.00	0.00	0.00
5	42	14.29	0.00	100.00	1.00	0.00	0.00	0.00	0.00	0.00	0.00
6	110	30.77	0.00	100.00	1.00	0.00	0.00	0.00	0.00	0.00	0.00
7	118	30.30	0.00	100.00	1.00	0.00	0.00	0.00	0.00	0.00	0.00
8	147	12.50	0.00	100.00	1.00	0.00	0.00	0.00	0.00	0.00	0.00
9	41	57.14	0.00	100.00	1.00	0.00	0.00	0.00	0.00	0.00	0.00
10	39	100.00	1.00	0.00	0.00	0.00	0.00	0.00	0.00	0.00	0.00
11	35	57.14	0.00	100.00	1.00	0.00	0.00	0.00	0.00	0.00	0.00
12	92	100.00	1.00	0.00	0.00	0.00	0.00	0.00	0.00	0.00	0.00
13	88	100.00	1.00	0.00	0.00	0.00	0.00	0.00	0.00	0.00	0.00
14	94	100.00	1.00	0.00	0.00	0.00	0.00	0.00	0.00	0.00	0.00

Applying pesticides regularly (yes = 1)

Cluster ID	Agents	Cumul (%)	Value	Cumul (%)	Value	Cumul (%)	Value	Cumul (%)	Value	Cumul (%)	Value
0	214	51.11	0.00	100.00	1.00	0.00	0.00	0.00	0.00	0.00	0.00
1	216	67.44	0.00	100.00	1.00	0.00	0.00	0.00	0.00	0.00	0.00
2	171	53.49	0.00	100.00	1.00	0.00	0.00	0.00	0.00	0.00	0.00
3	48	50.00	0.00	100.00	1.00	0.00	0.00	0.00	0.00	0.00	0.00
4	36	50.00	0.00	100.00	1.00	0.00	0.00	0.00	0.00	0.00	0.00
5	42	71.43	0.00	100.00	1.00	0.00	0.00	0.00	0.00	0.00	0.00
6	110	53.85	0.00	100.00	1.00	0.00	0.00	0.00	0.00	0.00	0.00
7	118	57.58	0.00	100.00	1.00	0.00	0.00	0.00	0.00	0.00	0.00
8	147	75.00	0.00	100.00	1.00	0.00	0.00	0.00	0.00	0.00	0.00
9	41	42.86	0.00	100.00	1.00	0.00	0.00	0.00	0.00	0.00	0.00
10	39	42.86	0.00	100.00	1.00	0.00	0.00	0.00	0.00	0.00	0.00
11	35	57.14	0.00	100.00	1.00	0.00	0.00	0.00	0.00	0.00	0.00
12	92	78.57	0.00	100.00	1.00	0.00	0.00	0.00	0.00	0.00	0.00
13	88	53.85	0.00	100.00	1.00	0.00	0.00	0.00	0.00	0.00	0.00
14	94	83.33	0.00	100.00	1.00	0.00	0.00	0.00	0.00	0.00	0.00

Liquidity per capita (1000 baht)

Cluster ID	Agents	Cumul (%)	Value	Cumul (%)	Value	Cumul (%)	Value	Cumul (%)	Value	Cumul (%)	Value
0	214	11.11	5.60	28.89	15.09	46.67	36.32	64.44	83.74	100.00	196.59
1	216	6.98	9.13	25.58	27.24	44.19	65.58	62.79	118.73	100.00	350.25
2	171	6.98	21.49	25.58	66.75	44.19	138.83	62.79	295.66	100.00	1191.75
3	48	12.50	13.83	25.00	18.75	50.00	30.53	75.00	40.95	100.00	59.96
4	36	16.67	20.98	33.33	39.92	50.00	47.79	66.67	68.95	100.00	151.55
5	42	14.29	23.74	28.57	32.98	42.86	59.18	71.43	137.22	100.00	436.92
6	110	7.69	10.50	23.08	27.44	38.46	45.81	69.23	107.69	100.00	404.23
7	118	3.03	7.69	27.27	27.02	51.52	68.32	75.76	129.28	100.00	325.97
8	147	16.67	18.91	33.33	45.31	50.00	89.17	66.67	179.67	100.00	344.91
9	41	14.29	66.02	28.57	81.52	42.86	100.17	71.43	129.25	100.00	228.34
10	39	14.29	68.72	28.57	70.21	42.86	90.61	71.43	168.71	100.00	285.47
11	35	14.29	113.47	28.57	189.04	42.86	268.94	71.43	304.23	100.00	421.79
12	92	14.29	26.87	28.57	55.01	42.86	83.90	71.43	106.36	100.00	138.98
13	88	7.69	44.74	23.08	60.97	38.46	106.71	69.23	146.52	100.00	316.07
14	94	16.67	141.00	33.33	172.23	50.00	198.97	66.67	246.66	100.00	436.68

Own motorbike (yes = 1)

Cluster ID	Agents	Cumul (%)	Value	Cumul (%)	Value	Cumul (%)	Value	Cumul (%)	Value	Cumul (%)	Value
0	214	28.89	0.00	100.00	1.00	0.00	0.00	0.00	0.00	0.00	0.00
1	216	18.60	0.00	100.00	1.00	0.00	0.00	0.00	0.00	0.00	0.00
2	171	20.93	0.00	100.00	1.00	0.00	0.00	0.00	0.00	0.00	0.00
3	48	37.50	0.00	100.00	1.00	0.00	0.00	0.00	0.00	0.00	0.00
4	36	33.33	0.00	100.00	1.00	0.00	0.00	0.00	0.00	0.00	0.00
5	42	28.57	0.00	100.00	1.00	0.00	0.00	0.00	0.00	0.00	0.00
6	110	23.08	0.00	100.00	1.00	0.00	0.00	0.00	0.00	0.00	0.00
7	118	12.12	0.00	100.00	1.00	0.00	0.00	0.00	0.00	0.00	0.00
8	147	4.17	0.00	100.00	1.00	0.00	0.00	0.00	0.00	0.00	0.00
9	41	42.86	0.00	100.00	1.00	0.00	0.00	0.00	0.00	0.00	0.00
10	39	100.00	1.00	0.00	0.00	0.00	0.00	0.00	0.00	0.00	0.00
11	35	100.00	1.00	0.00	0.00	0.00	0.00	0.00	0.00	0.00	0.00
12	92	21.43	0.00	100.00	1.00	0.00	0.00	0.00	0.00	0.00	0.00
13	88	15.38	0.00	100.00	1.00	0.00	0.00	0.00	0.00	0.00	0.00
14	94	16.67	0.00	100.00	1.00	0.00	0.00	0.00	0.00	0.00	0.00

Member of Royal Project (yes = 1)

Cluster ID	Agents	Cumul (%)	Value	Cumul (%)	Value	Cumul (%)	Value	Cumul (%)	Value	Cumul (%)	Value
0	214	93.33	0.00	100.00	1.00	0.00	0.00	0.00	0.00	0.00	0.00
1	216	86.05	0.00	100.00	1.00	0.00	0.00	0.00	0.00	0.00	0.00
2	171	93.02	0.00	100.00	1.00	0.00	0.00	0.00	0.00	0.00	0.00
3	48	87.50	0.00	100.00	1.00	0.00	0.00	0.00	0.00	0.00	0.00
4	36	16.67	0.00	100.00	1.00	0.00	0.00	0.00	0.00	0.00	0.00
5	42	28.57	0.00	100.00	1.00	0.00	0.00	0.00	0.00	0.00	0.00
6	110	46.15	0.00	100.00	1.00	0.00	0.00	0.00	0.00	0.00	0.00
7	118	36.36	0.00	100.00	1.00	0.00	0.00	0.00	0.00	0.00	0.00
8	147	45.83	0.00	100.00	1.00	0.00	0.00	0.00	0.00	0.00	0.00
9	41	100.00	0.00	0.00	0.00	0.00	0.00	0.00	0.00	0.00	0.00
10	39	57.14	0.00	100.00	1.00	0.00	0.00	0.00	0.00	0.00	0.00
11	35	42.86	0.00	100.00	1.00	0.00	0.00	0.00	0.00	0.00	0.00
12	92	57.14	0.00	100.00	1.00	0.00	0.00	0.00	0.00	0.00	0.00
13	88	30.77	0.00	100.00	1.00	0.00	0.00	0.00	0.00	0.00	0.00
14	94	16.67	0.00	100.00	1.00	0.00	0.00	0.00	0.00	0.00	0.00

Exposure – links to farmers who know about the innovation (#)

Cluster ID	Agents	Cumul (%)	Value	Cumul (%)	Value	Cumul (%)	Value	Cumul (%)	Value	Cumul (%)	Value
0	214	31.11	0.13	33.33	0.18	40.00	0.19	64.44	0.29	100.00	0.47
1	216	20.93	0.13	30.23	0.18	32.56	0.19	67.44	0.26	100.00	0.42
2	171	23.26	0.13	37.21	0.18	65.12	0.24	81.40	0.33	100.00	0.46
3	48	100.00	0.33	0.00	0.00	0.00	0.00	0.00	0.00	0.00	0.00
4	36	66.67	0.33	100.00	0.50	0.00	0.00	0.00	0.00	0.00	0.00
5	42	57.14	0.33	100.00	0.50	0.00	0.00	0.00	0.00	0.00	0.00
6	110	3.85	0.32	7.69	0.47	53.85	0.61	84.62	0.74	100.00	0.94
7	118	21.21	0.32	33.33	0.47	48.48	0.59	84.85	0.64	100.00	0.97
8	147	16.67	0.32	25.00	0.47	41.67	0.59	79.17	0.63	100.00	0.97
9	41	85.71	0.05	100.00	0.07	0.00	0.00	0.00	0.00	0.00	0.00
10	39	57.14	0.05	100.00	0.07	0.00	0.00	0.00	0.00	0.00	0.00
11	35	57.14	0.05	100.00	0.07	0.00	0.00	0.00	0.00	0.00	0.00
12	92	21.43	0.59	71.43	0.65	92.86	0.89	100.00	0.97	0.00	0.00
13	88	38.46	0.59	61.54	0.65	92.31	0.89	100.00	0.97	0.00	0.00
14	94	50.00	0.59	91.67	0.65	100.00	0.89	0.00	0.00	0.00	0.00

Village head (yes = 1)

Cluster ID	Agents	Cumul (%)	Value	Cumul (%)	Value	Cumul (%)	Value	Cumul (%)	Value	Cumul (%)	Value
0	214	100.00	0.00	0.00	0.00	0.00	0.00	0.00	0.00	0.00	0.00
1	216	100.00	0.00	0.00	0.00	0.00	0.00	0.00	0.00	0.00	0.00
2	171	95.35	0.00	100.00	1.00	0.00	0.00	0.00	0.00	0.00	0.00
3	48	100.00	0.00	0.00	0.00	0.00	0.00	0.00	0.00	0.00	0.00
4	36	100.00	0.00	0.00	0.00	0.00	0.00	0.00	0.00	0.00	0.00
5	42	85.71	0.00	100.00	1.00	0.00	0.00	0.00	0.00	0.00	0.00
6	110	88.46	0.00	100.00	1.00	0.00	0.00	0.00	0.00	0.00	0.00
7	118	93.94	0.00	100.00	1.00	0.00	0.00	0.00	0.00	0.00	0.00
8	147	79.17	0.00	100.00	1.00	0.00	0.00	0.00	0.00	0.00	0.00
9	41	100.00	0.00	0.00	0.00	0.00	0.00	0.00	0.00	0.00	0.00
10	39	100.00	0.00	0.00	0.00	0.00	0.00	0.00	0.00	0.00	0.00
11	35	100.00	0.00	0.00	0.00	0.00	0.00	0.00	0.00	0.00	0.00
12	92	100.00	0.00	0.00	0.00	0.00	0.00	0.00	0.00	0.00	0.00
13	88	100.00	0.00	0.00	0.00	0.00	0.00	0.00	0.00	0.00	0.00
14	94	100.00	0.00	0.00	0.00	0.00	0.00	0.00	0.00	0.00	0.00

Born in the area (yes = 1)

Cluster ID	Agents	Cumul (%)	Value	Cumul (%)	Value	Cumul (%)	Value	Cumul (%)	Value	Cumul (%)	Value
0	214	28.89	0.00	100.00	1.00	0.00	0.00	0.00	0.00	0.00	0.00
1	216	23.26	0.00	100.00	1.00	0.00	0.00	0.00	0.00	0.00	0.00
2	171	27.91	0.00	100.00	1.00	0.00	0.00	0.00	0.00	0.00	0.00
3	48	25.00	0.00	100.00	1.00	0.00	0.00	0.00	0.00	0.00	0.00
4	36	100.00	1.00	0.00	0.00	0.00	0.00	0.00	0.00	0.00	0.00
5	42	100.00	1.00	0.00	0.00	0.00	0.00	0.00	0.00	0.00	0.00
6	110	46.15	0.00	100.00	1.00	0.00	0.00	0.00	0.00	0.00	0.00
7	118	54.55	0.00	100.00	1.00	0.00	0.00	0.00	0.00	0.00	0.00
8	147	33.33	0.00	100.00	1.00	0.00	0.00	0.00	0.00	0.00	0.00
9	41	28.57	0.00	100.00	1.00	0.00	0.00	0.00	0.00	0.00	0.00
10	39	85.71	0.00	100.00	1.00	0.00	0.00	0.00	0.00	0.00	0.00
11	35	71.43	0.00	100.00	1.00	0.00	0.00	0.00	0.00	0.00	0.00
12	92	57.14	0.00	100.00	1.00	0.00	0.00	0.00	0.00	0.00	0.00
13	88	53.85	0.00	100.00	1.00	0.00	0.00	0.00	0.00	0.00	0.00
14	94	50.00	0.00	100.00	1.00	0.00	0.00	0.00	0.00	0.00	0.00

Annex V: Selected spatial inputs

Agent plots subdivided by 5 sectors (groups of villages)

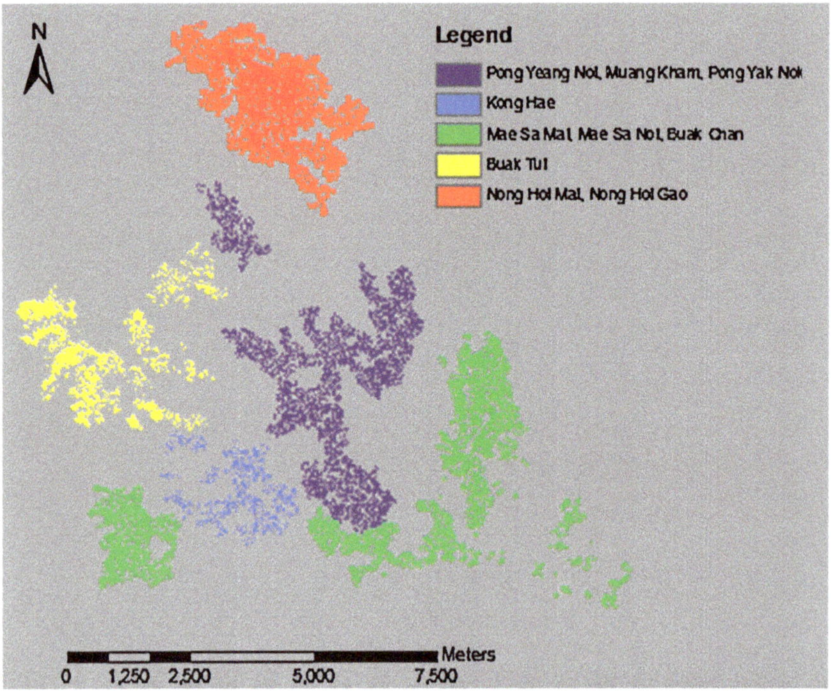

Agent farmsteads 168

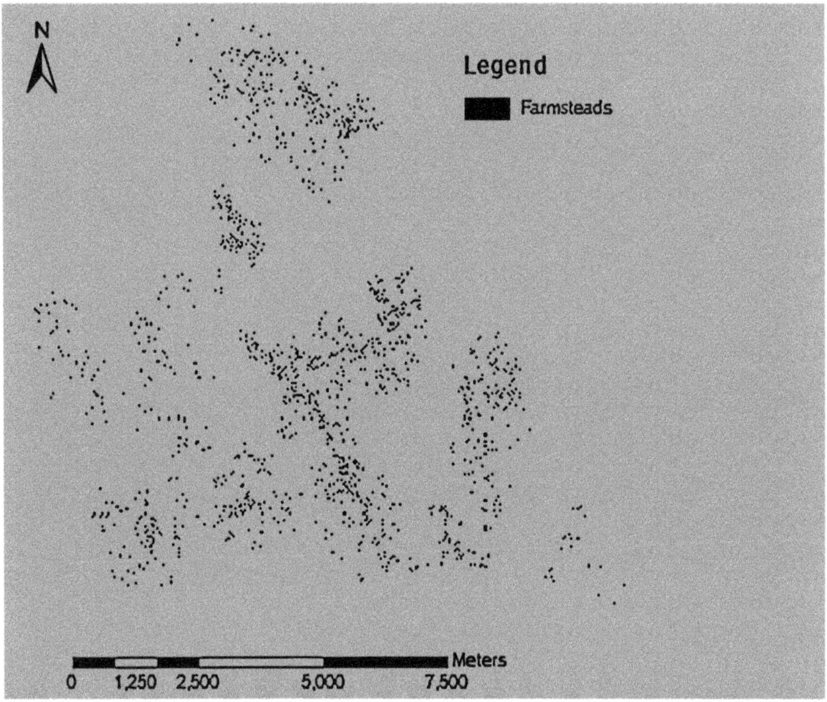

Development Economics and Policy

Series edited by Franz Heidhues, Joachim von Braun and Manfred Zeller

Band 1 Andrea Fadani: Agricultural Price Policy and Export and Food Production in Cameroon. A Farming Systems Analysis of Pricing Policies. The Case of Coffee-Based Farming Systems. 1999.

Band 2 Heike Michelsen: Auswirkungen der Währungsunion auf den Strukturanpassungsprozeß der Länder der afrikanischen Franc-Zone. 1995.

Band 3 Stephan Bea: Direktinvestitionen in Entwicklungsländern. Auswirkungen von Stabilisierungsmaßnahmen und Strukturreformen in Mexiko. 1995.

Band 4 Franz Heidhues / François Kamajou: Agricultural Policy Analysis – Proceedings of an International Seminar, held at the University of Dschang, Cameroon on May 26 and 27 1994, funded by the European Union under the Science and Technology Program (STD). 1996.

Band 5 Elke M. Förster: Protection or Liberalization? A Policy Analysis of the Korean Beef Sector. 1996.

Band 6 Gertrud Schrieder: The Role of Rural Finance for Food Security of the Poor in Cameroon. 1996.

Band 7 Nestor R. Ahoyo Adjovi: Economie des Systèmes de Production intégrant la Culture de Riz au Sud du Bénin: Potentialités, Contraintes et Perspectives. 1996.

Band 8 Jenny Müller: Income Distribution in the Agricultural Sector of Thailand. Empirical Analysis and Policy Options. 1996.

Band 9 Michael Brüntrup: Agricultural Price Policy and its Impact on Production, Income, Employment and the Adoption of Innovations. A Farming Systems Based Analysis of Cotton Policy in Northern Benin. 1997.

Band 10 Justin Bomda: Déterminants de l'Epargne et du Crédit, et leurs Implications pour le Développement du Système Financier Rural au Cameroun. 1998.

Band 11 John M. Msuya: Nutrition Improvement Projects in Tanzania: Implementation, Determinants of Performance, and Policy Implications. 1998.

Band 12 Andreas Neef: Auswirkungen von Bodenrechtswandel auf Ressourcennutzung und wirtschaftliches Verhalten von Kleinbauern in Niger und Benin. 1999.

Band 13 Susanna Wolf (ed.): The Future of EU-ACP Relations. 1999.

Band 14 Franz Heidhues / Gertrud Schrieder (eds.): Romania – Rural Finance in Transition Economies. 2000.

Band 15 Katinka Weinberger: Women's Participation. An Economic Analysis in Rural Chad and Pakistan. 2000.

Band 16 Christof Batzlen: Migration and Economic Development. Remittances and Investments in South Asia: A Case Study of Pakistan. 2000.

Band 17 Matin Qaim: Potential Impacts of Crop Biotechnology in Developing Countries. 2000.

Band 18 Jean Senahoun: Programmes d'ajustement structurel, sécurité alimentaire et durabilité agricole. Une approche d'analyse intégrée, appliquée au Bénin. 2001.

Band 19 Torsten Feldbrügge: Economics of Emergency Relief Management in Developing Countries. With Case Studies on Food Relief in Angola and Mozambique. 2001.

Band 20 Claudia Ringler: Optimal Allocation and Use of Water Resources in the Mekong River Basin: Multi-Country and Intersectoral Analyses. 2001.

Band 21 Arnim Kuhn: Handelskosten und regionale (Des-)Integration. Russlands Agrarmärkte in der Transformation. 2001.

Band 22 Ortrun Anne Gronski: Stock Markets and Economic Growth. Evidence from South Africa. 2001.

Band 23 Patrick Webb / Katinka Weinberger (eds.): Women Farmers. Enhancing Rights, Recognition and Productivity. 2001.

Band 24 Mingzhi Sheng: Lebensmittelkonsum und -konsumtrends in China. Eine empirische Analyse auf der Basis ökonometrischer Nachfragemodelle. 2002.

Band 25 Maria Iskandarani: Economics of Household Water Security in Jordan. 2002.

Band 26 Romeo Bertolini: Telecommunication Services in Sub-Saharan Africa. An Analysis of Access and Use in the Southern Volta Region in Ghana. 2002.

Band 27 Dietrich Müller-Falcke: Use and Impact of Information and Communication Technologies in Developing Countries' Small Businesses. Evidence from Indian Small Scale Industry. 2002.

Band 28 Wolfram Erhardt: Financial Markets for Small Enterprises in Urban and Rural Northern Thailand. Empirical Analysis on the Demand for and Supply of Financial Services, with Particular Emphasis on the Determinants of Credit Access and Borrower Transaction Costs. 2002.

Band 29 Wensheng Wang: The Impact of Information and Communication Technologies on Farm Households in China. 2002.

Band 30 Shyamal K. Chowdhury: Institutional and Welfare Aspects of the Provision and Use of Information and Communication Technologies in the Rural Areas of Bangladesh and Peru. 2002.

Band 31 Annette Luibrand: Transition in Vietnam. Impact of the Rural Reform Process on an Ethnic Minority. 2002.

Band 32 Felix Ankomah Asante: Economic Analysis of Decentralisation in Rural Ghana. 2003.

Band 33 Chodechai Suwanaporn: Determinants of Bank Lending in Thailand: An Empirical Examination for the Years 1992 to 1996. 2003.

Band 34 Abay Asfaw: Costs of Illness, Demand for Medical Care, and the Prospect of Community Health Insurance Schemes in the Rural Areas of Ethiopia. 2003.

Band 35 Gi-Soon Song: The Impact of Information and Communication Technologies (ICTs) on Rural Households. A Holistic Approach Applied to the Case of Lao People's Democratic Re- public. 2003.

Band 36 Daniela Lohlein: An Economic Analysis of Public Good Provision in Rural Russia. The Case of Education and Health Care. 2003.

Band 37 Johannes Woelcke. Bio-Economics of Sustainable Land Management in Uganda. 2003.

Band 38 Susanne M. Ziemek: The Economics of Volunteer Labor Supply. An Application to Countries of a Different Development Level. 2003.

Band 39 Doris Wiesmann: An International Nutrition Index. Concept and Analyses of Food Insecurity and Undernutrition at Country Levels. 2004.

Band 40 Isaac Osei-Akoto: The Economics of Rural Health Insurance. The Effects of Formal and Informal Risk-Sharing Schemes in Ghana. 2004.

Band 41 Yuansheng Jiang: Health Insurance Demand and Health Risk Management in Rural China. 2004.

Band 42 Roukayatou Zimmermann: Biotechnology and Value-added Traits in Food Crops: Relevance for Developing Countries and Economic Analyses. 2004.

Band 43 F. Markus Kaiser: Incentives in Community-based Health Insurance Schemes. 2004.

Band 44 Thomas Herzfeld: *Corruption begets Corruption.* Zur Dynamik und Persistenz der Korruption. 2004.

Band 45 Edilegnaw Wale Zegeye: The Economics of On-Farm Conservation of Crop Diversity in Ethiopia: Incentives, Attribute Preferences and Opportunity Costs of Maintaining Local Varieties of Crops. 2004.

Band 46 Adama Konseiga: Regional Integration Beyond the Traditional Trade Benefits: Labor Mobility contribution. The Case of Burkina Faso and Côte d'Ivoire. 2005.

Band 47 Beyene Tadesse Ferenji: The Impact of Policy Reform and Institutional Transformation on Agricultural Performance. An Economic Study of Ethiopian Agriculture. 2005.

Band 48 Sabine Daude: Agricultural Trade Liberalization in the WTO and Its Poverty Implications. A Study of Rural Households in Northern Vietnam. 2005.

Band 49 Kadir Osman Gyasi: Determinants of Success of Collective Action on Local Commons. An Empirical Analysis of Community-Based Irrigation Management in Northern Ghana. 2005.

Band 50 Borbala E. Balint: Determinants of Commercial Orientation and Sustainability of Agricultural Production of the Individual Farms in Romania. 2006.

Band 51 Pamela Marinda: Effects of Gender Inequality in Resource Ownership and Access on Household Welfare and Food Security in Kenya. A Case Study of West Pokot District. 2006.

Band 52 Charles Palmer: The Outcomes and their Determinants from Community-Company Contracting over Forest Use in Post-Decentralization Indonesia. 2006.

Band 53 Hardwick Tchale: Agricultural Policy and Soil Fertility Management in the Maize-based Smallholder Farming System in Malawi. 2006.

Band 54 John Kedi Mduma: Rural Off-Farm Employment and its Effects on Adoption of Labor Intensive Soil Conserving Measures in Tanzania. 2006.

Band 55 Mareike Meyn: The Impact of EU Free Trade Agreements on Economic Development and Regional Integration in Southern Africa. The Example of EU-SACU Trade Relations. 2006.

Band 56 Clemens Breisinger: Modelling Infrastructure Investments, Growth and Poverty Impact. A Two-Region Computable General Equilibrium Perspective on Vietnam. 2006.

Band 57 Meike Wollni: Coping with the Coffee Crisis. An Analysis of the Production and Marketing Performance of Coffee Farmers in Costa Rica. 2007.

Band 58 Franklin Simtowe: Performance and Impact of Microfinance. Evidence from Joint Liability Lending Programs in Malawi. 2008.

Band 59 Xiangping Jia: Credit Rationing and Institutional Constraint. Evidence from Rural China. 2008.

Band 60 Holger Seebens: The Economics of Gender and the Household in Developing Countries. 2009.

Band 61 Stephan Piotrowski: Land Property Rights and Natural Resource Use. An Analysis of Household Behavior in Rural China. 2009.

Band 62 Sebastian M. Scholz: Rural Development through Carbon Finance. Forestry Projects under the Clean Development Mechanism of the Kyoto Protocol. Assessing Smallholder Participation by Structural Equation Modeling. 2009.

Band 63 Jakob Rupert Friederichsen: Opening Up Knowledge Production through Participatory Research? Agricultural Research for Vietnam's Northern Uplands. 2009.

Band 64 Olivier Ecker: Economics of Micronutrient Malnutrition. The Demand for Nutrients in Sub-Saharan Africa. 2009.

Band 65 Julia Johannsen: Operational Assessment of Monetary Poverty by Proxy Means Tests. 2009

Band 66 Ephraim Nkonya / Nicolas Gerber / Philipp Baumgartner / Joachim von Braun / Alessandro De Pinto / Valerie Graw / Edward Kato / Julia Kloos / Teresa Walter: The Economics of Land Degradation. Toward an Integrated Global Assessment. 2011.

Band 67 S. Idriss Nazaire Houssou: Operational Poverty Targeting by Proxy Means Tests. Models and Policy Simulations for Malawi. 2013.

Band 68 Abdul Salam Lodhi: Education, Child Labor and Human Capital Formation in Selected Urban and Rural Settings of Pakistan. 2013.

Band 69 Evita Hanie Pangaribowo: Household Food Consumption, Women´s Asset and Food Policy in Indonesia. 2013.

Band 70 Dan Liu: China's New Rural Cooperative Medical Scheme. Evolution, Design and Impacts. 2013.

Band 71 Camille Saint-Macary: Microeconomic Impacts of Institutional Change in Vietnam's Northern Uplands. Empirical Studies on Social Capital, Land and Credit Institutions. 2014.

Band 72 Beatrice Wambui Muriithi: Commercialization of Smallholder Horticultural Farming in Kenya. Poverty, Gender, and Institutional Arrangements. 2014.

Band 73 Christian C. W. Grovermann: Assessment of Pesticide Use Reduction Strategies for Thai Highland Agriculture. Combining Econometrics and Agent-based Modelling. 2015.

www.ingramcontent.com/pod-product-compliance
Ingram Content Group UK Ltd.
Pitfield, Milton Keynes, MK11 3LW, UK
UKHW022222051026
15624UKWH00002B/300

9 783631 657843